FLOWING
THROUGH
TIME

FLOWING THROUGH TIME

A History of the Iowa Institute of Hydraulic Research

By Cornelia F. Mutel

With a Foreword by Virendra C. Patel

Iowa Institute of Hydraulic Research
Iowa City, Iowa

Iowa Institute of Hydraulic Research, College of Engineering,
University of Iowa, Iowa City, Iowa, 52242
Copyright © 1998 by the Iowa Institute of Hydraulic Research
Printed in the United States of America
First edition, 1998

The preparation of this book was assisted by a grant from the University of
Iowa's Cultural Affairs Council.

Library of Congress Cataloging-in-Publication Data
Mutel, Cornelia Fleischer.
    Flowing through time: a history of the Iowa Institute of Hydraulic
    Research/by Cornelia F. Mutel: with a foreword by Virendra C. Patel.—1st ed.
        p.   cm.
    Includes bibliographical references and index.
    ISBN 0-87414-108-7 (pbk.: alk. paper)
    1. Iowa Institute of Hydraulic Research—History. 2. Hydraulic
    engineering—Research—History. 3. Hydraulics—History. 4. Hydraulic
    engineering—United States—History. 5. University of Iowa—History I. Title.
TC158.M88   1998
627'.0720777'655—dc21

Contents

Foreword

Speculating on the future of the Iowa Institute of Hydraulic Research (IIHR), my predecessor John F. Kennedy used to pose the rhetorical question, "If you cannot succeed in the water business, in what business can you succeed?" He would then comment on the good fortune of our being active in research involving all four elements of the physical universe—water, air, fire, and earth—reasoning that due to the pervasive importance of water and its interaction with the remaining three elements, our future was assured because our work would never be finished. Well, this book by Cornelia F. Mutel gives substance to those words in a manner that is at once coherent and congenial. It is a book about research on problems related to water and fluid mechanics in general, and how that research was influenced by socioeconomic currents in the nation.

The history of science, and particularly that of big science, is quite fascinating and has captured the imagination of readers and moviegoers alike. But this book is not a history of big science. It is the story of a hydraulics laboratory situated in a campus town in the middle of the U.S., in a state located between two great rivers but far from either coast. What makes this story worth telling? Of all the institutions created by humans, the university stands out as the finest and most enduring creation. This is so because universities are constantly replenished by new people with new ideas, coming to learn and teach and challenge one another to rise to higher levels of understanding and problem solving. Such is also the story of IIHR, because it is closely linked with the academic departments of the College of Engineering that are engaged in fluids engineering education. In addition, as opportunity presents itself, IIHR works with other units of the university such as internal medicine, urology, radiology, physiology, and physical education, on fluids-related problems of mutual interest. The institute's joint mission of teaching and performing research has benefited from the constant influx of people and ideas.

From its inception more than seven decades ago, the institute has attracted students and staff from around the world. Thus, the first letter in IIHR could stand for "international." In fact, because of this international cast, the institute has no globalization plans—it has been global all along. Today we have students from 10 countries and staff with roots in 11 countries. The attraction to Iowa is certainly not the weather. The fact that hydraulicians, hydrodynamicists, and fluids engineers of the caliber of Nagler, Rouse, Landweber, and Kennedy, from different backgrounds and with different passions, have worked here is a testimony to the vitality of the organization. The institute has attracted and, more important, has retained people trained in diverse disciplines, who elsewhere would work in quite different academic departments confined by their traditional divides. Today, people with degrees in aerospace, civil, computer, electrical, and mechanical engineering, and in naval architecture, geology, geography, physics, and mathematics work under one roof, collaborating on projects in ways that would be quite difficult to accomplish at other institutions. This broad-ranging activity of teaching and research in a single organization has been the hallmark of the institute, one that has empowered it over the years to flow with the times and respond quickly, resourcefully, and effectively to new opportunities in research. This is also an engineering research organization that does not allow the usual academic debate about basic versus applied research to sway it too far from tackling and solving practical, real-world problems of the time. This mix and blend of the institute's engagement over the years is vividly brought out in the book.

Visitors to the institute, awed by the diversity of facilities and projects that they see, frequently ask questions such as "Why a ship-towing tank in Iowa?" or "Why do you have so many wind tunnels when you do not offer degrees in aeronautics?" Such questions are best answered by reference to the history of the institute, and as Mutel tells that history, it is the people who came, stayed, and worked here who made it all possible. The unifying theme in this apparent diversity is the parent field of fluid mechanics, which engages the attention of almost all branches of engineering in addition to physics and mathematics.

The collective effort and vision of the institute's staff has created and sustained an organization that many seek to emulate but few come to match. The institute remains a vital and vibrant place where students, teachers, and researchers collectively address problems of fluids engineering. It has, over the years, managed to be on the leading edge of education and research. Today it seems to be well positioned for the future as society becomes ever more aware of the interactions among the four elements of nature, and as the current buzzwords of multidisciplinary research and globalization of education find concrete expression in the work that is conducted here. In this respect, the institute's business—the water business—seems secure.

Virendra C. Patel
Director, Iowa Institute of Hydraulic Research

 **Preface**

When I first started working on this book, I conceived of it as a series of vignettes supplemented by photographs. I was not interested in writing a dry chronology of events, nor did I wish to write a technical history that would appeal only to hydraulicians or engineers. I knew that the Iowa Institute of Hydraulic Research had performed research crucial to the workings of modern society. Why not, I thought, select a few stories of such impact for each decade as a way to present this fascinating and relevant history? Such stories would be sure to interest and inform a broader audience that included persons with little or no training in engineering or science as well as those with connections to hydraulics or IIHR.

This approach remained operable for the first few months of the two years required to complete the book. Then the activities and personalities that have energized this organization since 1920 took hold, and I realized that I needed to let go of my preconceptions and "flow through time" along with them, as the book's title suggests. As I did so, the passion that has characterized IIHR directors started to flow through my veins also. I became increasingly absorbed in the complexity and impact of research initiatives that have been completed here. These personalities and IIHR's many research efforts needed to be thoroughly summarized, I felt, if IIHR's history were to be granted its full worth. Yet I wanted to retain the approach of a storyteller, enticing outsiders into the book by entertaining them with IIHR's decades of evolving activities. Through this latter approach, I felt that I could educate my readers surreptitiously about properties of the liquid that is most crucial to the lives of all of us: water.

This evolving mixture of goals explains the book's hybrid structure. I have divided the book into three main sections, each of which examines about a third of the century. IIHR's history falls amazingly neatly into these three time categories, with changes in directorship (and the accompanying changes in research emphasis) coinciding with major construction and technological advances.

The first chapter in each section (chapters 2, 6, and 12) includes a description of the activities of a long-term director of IIHR. (Chapters 2 and 12 also include information on shorter-term directors.) The second chapter of each section (chapters 3, 7, and 13) presents the evolving research picture throughout the years in question. The third chapter considers logistical developments—the practical structures or technologies that marked the period in question: first the construction of the Hydraulics Laboratory (chapter 4), then the art of physical modeling and the development of IIHR's shop (chapter 8), and then the increasing importance of computers (chapter 14). The information in these latter three chapters is not limited to the respective sections in which they are included. The construction and the impact of physical models, for example, have been integral components of IIHR's entire history, even though this topic is examined primarily in section II of the book.

In each of the three sections, I have told "IIHR stories"—my more detailed descriptions of a particular research initiative, activity, or period. These descriptions, which fill chapters 5, 9–11, and 15–20, look at a single slice of IIHR's life and researchers in a manner that explains their broader importance and interest to the reader.

Due to the complexity and longevity of the institute, I had to be selective about the research projects and the people I discussed. The emphasis on IIHR's directors, for example, each of whom has molded the research program with a strong hand, might incorrectly imply that they alone have shaped the institute's direction. This emphasis is not meant to underplay the importance of IIHR's research engineers, other staff members, and students. All of these people form the true body and spirit of IIHR. They all performed the research, wrote the papers and books, and (when faculty members) educated the students who spread IIHR's influence throughout the world. While I have attempted to include each research engineer by name, it was impossible to do so for all of the many other staff members, post-doctoral associates, visiting scientists, and the like. My omissions of specific reference to these individuals, inadvertent omission of research engineers, and failure to note the breadth of effort of IIHR's many researchers are not intended to negate the importance of their contributions.

This book fits into a tradition of several earlier IIHR histories, the last of which was published in 1971. In keeping with those histories, I have included a list of staff members and of graduate-degree recipients (along with their thesis titles and advisors) in two appendices.

The book is based on information gleaned from IIHR's bulletins and other publications, unpublished documents held in IIHR's archives, and documents held in the University of Iowa Archives (in the Special Collections Department at the University Libraries), with additional information provided through interviews of IIHR staff members, selected former employees, and sometimes their families. I attempted to ensure accuracy by meticulous attention to detail when working with archives, by cross-referencing sources and eliminating information not corroborated by multiple sources, and by soliciting reviews of multiple drafts of each chapter from current and former IIHR staff members, in particular those chapters that discussed the staff member's research efforts or field of expertise.

To improve readability and more fully engross the reader, I purposefully have employed the present verb tense (a practice often frowned upon in historical writing). I also have eliminated references to specific sources of information and to quotes. However, an annotated copy of the manuscript (with references to source material) is available in the IIHR Reference Room for those seeking more information.

All photographs, unless otherwise credited, are from IIHR researchers or IIHR archives. All photos are reproduced with kind permission from their owners. The location of specific photos has been noted in the annotated manuscript.

An edited version of chapter 9 was included in the *Iowa Alumni Quarterly* (Winter 1997), and an article based on a compilation of chapters 2–5 was published in *Iowa Heritage Illustrated* (winter 1996). An excerpt from chapter 10 was printed in the University of Iowa's *Illumine* (volume 1, issue 1, Spring 1998).

A subject this complex depends on help from many others. I have talked with nearly all IIHR staff members in the course of the project, asking them to provide or clarify information, review portions of the manuscript, or furnish photographs or other assistance, and they have responded graciously and cheerfully to my interruptions. My gratitude to them is not diminished because I am unable to list all who have assisted me on a regular basis.

Perhaps most crucial to the book's completion was Rob Ettema, who talked with me almost every week throughout the project about the book's material and its shaping. He also read and commented on the entire manuscript, as did V. C. Patel and Marlene Janssen. The book could not have been completed without Patel's support of and belief in the project and his allocation of institute funding, nor could I have proceeded through the more difficult chapters without Janssen's welcoming open door and constant encouragement.

This book was partially funded by a grant from the University of Iowa's Cultural Affairs Council.

Many associated with IIHR but not currently on its staff provided assistance. Phil Hubbard discussed and reviewed materials in section II. Luceille Howe and Dale and Ella Harris also tuned me in to the spirit of those earlier times. Robert Nagler, son of Floyd Nagler, and Allan H. Rouse, son of Hunter Rouse and trustee of the Rouse Trust, met with me to discuss their fathers and opened family archives to me. Edwin Rood helped me understand IIHR's contributions to ship hydrodynamics and reviewed the chapter on this subject. Allen Chwang, Jim Cramer, Simon Ince, and others provided helpful insights.

Dan Daly, in charge of IIHR's Reference Room, was invaluable in locating photographs and digging out obscure information. He regularly alerted me to materials and sources that I may have overlooked. Earl Rogers, University Archivist, and others in Special Collections also helped me locate information.

Thanks to the several contributors of photographs indicated in the photo credits and to Mike Kundert for preparing the map of IIHR buildings. Twila Meder prepared both appendices.

The book was designed by Parker Davis Graphics. Patti O'Neill, of the university's Printing Department, designed the cover.

Prasenjit Gupta and Linda Fisher assisted with final manuscript preparation.

This book could not have been completed without the assistance of each of the above-mentioned persons. Equally valuable to me, as author and project coordinator, were the many smiles, encouraging comments, and expressions of interest in the project, which lightened my days and gave wings to my efforts. Many, many thanks to all, for each and every one of these contributions.

Despite all this assistance, a discourse on an institution as diverse and long-lived as IIHR cannot help but contain errors of omission and misinterpretation. I take full responsibility for any such mistakes and oversights.

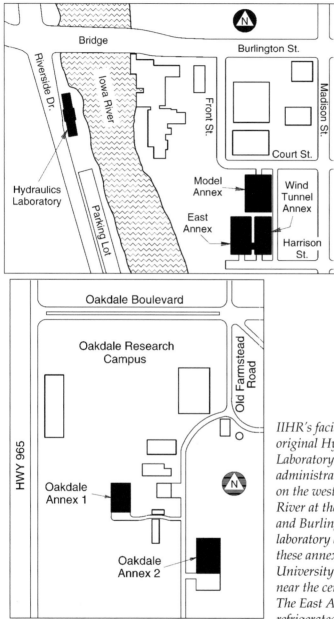

IIHR's facilities include the original Hydraulics Laboratory (now IIHR's administrative center), located on the west side of the Iowa River at the corner of Riverside and Burlington, and five laboratory annexes. Three of these annexes are on the University of Iowa campus near the center of Iowa City. The East Annex contains three refrigerated laboratories, a large environmental flume, and the mechanical shop. The Model Annex houses a constantly changing array of hydraulic models that are built according to current research needs. The Wind Tunnel Annex, in addition to several wind tunnels and hydrometeorology laboratories and equipment, also holds offices and a computer laboratory. The two annexes on the university's Oakdale Campus, six miles northwest of the Hydraulics Laboratory, provide unobstructed open space for very large models.

1. The Iowa Institute of Hydraulic Research: An Introduction

The Hydraulics Laboratory rises as an imposing red-brick edifice on one of the busiest corners of Iowa City, Iowa. Near the downtown area on the University of Iowa campus, with one side fronting Riverside Drive, a second facing Burlington Street, and a third abutting the Iowa River, this building is seen by thousands every day. Many might think of it as a typical academic "ivory tower," if they think about it at all. Some wonder if it is part of the university's nearby power plant. A few might wonder if the name "Hydraulics Laboratory," etched in stone above the central door, means that this is part of the city's water-treatment plant. Those entering the building and taking a cursory glance at their surroundings are not likely to be impressed: to their right they will see a pre–World War II pipe assembly and flume and a dated elevator that carries people and mail to the central offices on the fourth floor; to their left lies a Reference Room with overflowing shelves. Only the initiated, those who are hydraulicians or fluid mechanicians with some inside association, are likely to realize that this building's walls, if they could talk, would tell the ageless story of humans attempting to understand and control that most basic of the earth's constituents, water. Here the tale of humans interacting with their natural world, trying to decipher its components, trying to control its destructive forces, and trying to harvest its bounty, is told day after day, year after year, each time with a slightly different emphasis and slant.

Careful observers taking a closer look at the building might start to see signs of the passion and complexity of this search for truths about our elemental world. Entering the front door, two millstones are embedded along the staircase: someone here cared not just about the present uses of water, but also studied its use in the past. A fluids teaching lab occupies the first floor of the building: the training of students must be central to those working here. As one gains access to one of the research laboratories, high-tech research equipment comes into view. In one laboratory, a Plexiglas-sided tank of water for generating and observing vortices stands next to a laser and particle image velocimeter for measuring the velocity field in hydraulic pump intakes. In another lab, other laser-based equipment assesses fluid movement around a model ship hull traversing a towing tank. It's evident that researchers here investigate both fundamental processes and their applications to real-world situations. One's awareness of the diversity of studies multiplies as one passes offices occupied by hydrometeorologists, river hydraulicians, and environmental engineers, or as one peeks over a shoulder and witnesses a student accessing a supercomputer to develop a computational model that replicates one of the many physical models located in nearby laboratory annexes.

These are some of the many facets of the Iowa Institute of Hydraulic Research (IIHR), one of the world's preeminent institutes for investigating the diverse range of fluid-mechanics processes associated with water flow, be it in rivers, as moisture in the air, or as flow around and within diverse structures. Within the walls of the Hydraulics Laboratory and its five associated annexes, approximately a hundred research engineers, graduate students, and various technical and administrative-support staff struggle to answer multifarious questions related to water's behavior, its flow, and its transport of diverse materials. The struggle began in 1920 with the opening of the first Hydraulics Laboratory. In the intervening years, questions have been answered only to pose new enigmas. Processes have been deciphered in increasingly more detailed and complex manners. And the course and effects of water's flow through the modern world have continually been recast.

In the process, workers here have influenced how the Mississippi River has been reshaped, how engineers in China have designed their dams, how students in Venezuela have been taught fluid mechanics, and how fish could overcome the boundaries created by modern hydropower plants. Researchers have increased our appreciation for the cogitations of Leonardo da Vinci, and assembled collections of rare books, and struggled to improve our

IIHR's Hydraulics Laboratory

comprehension of water's flow through the atmosphere and our understanding of the generation of turbulence. Investigators have attacked the same basic problems through an increasingly more diverse set of approaches, and also have embraced emerging theories and newly perceived problems posed by fluid flows.

These mental explorations mark the lives of researchers who have devoted themselves intensely to their work. This is not to say that the course of IIHR has always been smooth. Research protocols have adapted to the needs of the times as well as the researchers' desires. Yet a study of IIHR's history reveals an amazing number of serendipitous conjunctions. The first director, Floyd Nagler, became good friends with a U.S. Department of Agriculture colleague who commenced an inflow of government funds that carried IIHR through the 1930s. A new director with a passion for fundamental fluid mechanics, Hunter Rouse, took the reins as the U.S. government expanded funding of theoretical research. He handed his office over to his successor John Kennedy, who loved defusing industrial and environmental problems, just a few years before environmental concerns and the energy crisis shunted funding into related endeavors and power-plant research.

At times IIHR seemed as if it might falter. It struggled, for example, to maintain financial viability during the Great Depression, along with the Civilian Conservation Corps workers who worked here then. Yet throughout its long history, IIHR has bucked the stream and managed to rise to the top, surviving when other hydraulics laboratories were faltering, growing and maintaining a complex mixture of efforts when other labs were becoming more restricted. The efforts of those working here have flowed like a braided stream, with rivulets sometimes crossing and merging, sometimes separating, but all the while flowing forward as a group in a complex manner.

In the 1990s, IIHR's research remains as diverse as it ever has been. What characteristics describe this relatively large, multidisciplinary laboratory? For one thing, even though IIHR has always been largely self-supporting and remains administratively distinct, it is not an insular unit. The institute has always been a unit within the University of Iowa's College of Engineering. Most IIHR research engineers hold joint appointments in the college, either in the Department of Civil and Environmental Engineering or in Mechanical Engineering. In earlier years, when the college was structured otherwise, appointments were joint with the Division of Energy Engineering, and before 1974 they were held with the Department of Mechanics and Hydraulics. Research efforts at IIHR are thus balanced with teaching activities and the mentoring of graduate students.

IIHR's staff has regularly contributed to these larger units in other ways also, for example by chairing departments and serving on various committees. One IIHR director, Hunter Rouse, became dean of the Engineering College. Another researcher, Phil Hubbard, stepped into the university's administration as dean of Academic Affairs. The institute, because of its renown, has regularly been recognized as one of the college's exemplary units with a national and international reputation, and as such contributes stature and stability to the college. As early as 1938, IIHR was acclaimed as the pride of the University of Iowa's Engineering College. Then, as later, its reputation became a strong reason for maintaining the college during the perennial attempts to merge this college with that of Iowa State University.

IIHR can be characterized by its diversity, some of which emerges from the mixing of civil and environmental engineers with mechanical engineers and researchers from other disciplines. This diversity of training and approaches is complemented by a blend of fundamental and applied studies and by a variety of research techniques, with IIHR claiming significant experience in both physical and numerical modeling efforts. Each of these approaches requires specialized expertise and significant financial investment. Thus, most hydraulics laboratories emphasize one or the other. IIHR is one of the few where both approaches remain highly developed and integrated.

The mixing of people, disciplines, and research approaches allows the daily interchange of ideas as well as techniques. Equipment from fundamental studies may be pulled by other observant researchers into applied studies, or vice versa. Physical models are used to validate numerical models, which in turn are used to design physical-model studies. Such synergism requires a laboratory of the size and diversity of IIHR. The rich mixture has benefited IIHR in another way: by increasing its ability to adjust to changes in funding, and thus to struggle through periods when funding protocols were changing and a more unidimensional laboratory with less resiliency might have failed.

IIHR's diversity is likewise fed by the constant influx of students and visitors from other countries, a migration that has held true throughout IIHR's history. The institute's international stature soared during Rouse's tenure, when his texts and visits abroad brought students and visitors here seeking classes with the "master." IIHR's international attraction and its networking have been maintained. However, the flow goes both ways, with IIHR researchers consulting, traveling, and teaching abroad, and participating in international collaborations. Outreach activities also have always included the organizing of conferences and major contributions to the publication of professional literature, both directly and through editorial roles. The wealth of international connections has sometimes prompted the claim that IIHR is better known abroad than in its own homeland.

These multiple factors and characteristics of IIHR, like those of all life, continue to evolve. IIHR never has been static. The protocols set today will be revised tomorrow to address new problems and questions. Such has been the story of IIHR since its earliest days. Such is the story told in this book.

IIHR staff, 1932

I UNCHARTED RIVERS
The Early Years
1920–1933

2. Floyd Nagler
and the Birth of IIHR

It is said that Floyd Nagler was a man of boundless energy and drive, capable of balancing ten activities simultaneously and keeping them all going. That in summer he often rose at 4:30 a.m. to tend the large family garden, because he believed that fresh vegetables and fruits were crucial to his family's health, and then walked swiftly off to an early start at the Hydraulics Laboratory. That at night he carried his love of water back home with him, damming a small watercourse to build his children a pond for swimming and fishing. That he approached his prolific efforts at the university with eagerness, efficiency, precision, and thoroughness, inspiring the same in others, who then had trouble keeping up with him. Not the least of his tasks before his premature death at age 41 was the establishment of IIHR.

Perhaps it is to be expected that Nagler was passionate and intense in his efforts. He was, after all, the son of an avid Methodist minister who had been fervent in conducting revival meetings and promoting prohibition. True to his family calling, the adult Floyd Nagler took over the pulpit whenever the Methodist minister of his Iowa City church left town. His elder son, Robert, states that Floyd's major purpose in life was the reconciliation of science, technology, and religion: he saw scientists as persons capable of explaining the Bible and the workings of God's creation to the lay public.

Nagler was born on January 11, 1892, and raised in Michigan. He graduated in 1914 from Michigan State College in East Lansing with a bachelor's degree in civil engineering, and within the following three years received master's and doctoral degrees in engineering from the University of Michigan. After short stints in the military and in an engineering consulting firm, he joined the State University of Iowa faculty as assistant professor of Mechanics and Hydraulics in September 1920. He proceeded up the ranks, becoming a professor of Hydraulic Engineering in 1927 and receiving the American Society of Civil Engineers' Collingwood Prize for Juniors twice (in 1919 and 1920), the Norman Medal (with Albion Davis) in 1931, and the J. James R. Croes Medal (with David Yarnell) in 1932. He was elected president of ASCE's Iowa Section the year preceding his death.

Nagler's primary legacy was the establishment of the physical and organizational structure of the Iowa Institute of Hydraulic Research. He had been brought to Iowa to foster the research of a tiny laboratory with under 500 square feet of floor space; he left a five-storied laboratory well over 50 times as large. He guided the transition of this Hydraulics Laboratory to the full-blown IIHR and served as the founding director of both the laboratory and the institute. During his 13-year tenure, his staff grew from 4 (Nagler himself, a mechanic, and two part-time assistants) to 26. Nineteen of these 26 were full-time employees, and 15 were engineers.

Nagler did far more than build a structure. As the first to steer the vessel, he set IIHR on its course and established the traditions that were to become its hallmark: traditions of high productivity, excellence and innovation in research, and service not only to the growing field of hydraulics but also to the many organizations and agencies that brought their questions and problems in search of solutions. In doing so, he established IIHR as one of the premier hydraulics laboratories in the country and a stellar unit of the University of Iowa and its engineering college.

The Hydraulics Laboratory's standards of excellence also were transmitted to the more than 50 graduate students who trained under Nagler and then carried their ideals and knowledge to other institutions and regions, where they applied them afresh to new situations. Nagler's commitment to students was reflected in his many comments on how he guided and counseled his protégés, and in his dedication to providing ample hands-on learning and research experience.

These qualities led to an amazing proliferation of research that exceeded even Nagler's expectations. At first, research work at the Hydraulics Laboratory was synonymous with work completed by

Floyd Nagler, IIHR's founder and first director, was a
vigorous and intense man, passionately dedicated to his
institute, his research (such as this 1924 effort to measure the
Mississippi's flow at the Keokuk Dam), and his many
personal interests.

Nagler; he was, after all, the only researcher present. As his staff
grew, Nagler remained involved in the Hydraulics Laboratory's research as well as its administration, and he seemed to have a thorough grasp of IIHR's diverse efforts and their import. His own efforts in later years focused on surveys regarding the water resource

potential and flow characteristics of the Mississippi, its immediate tributaries, and small rivers in Iowa, performed for the U.S. Army Corps of Engineers and for Iowa's Fish and Game Commission and Board of Conservation. These efforts are discussed in more depth in chapter 5.

Although the 41-year-old Nagler had every right to think that he would be present to guide his beloved IIHR for several more decades, his sudden and unexpected death left IIHR on an upward trajectory. While this course was set in part by the societal needs of the time, Nagler's dedication, personality, and talents could not help but complement the thrust. He was praised for his wealth and breadth of knowledge, which combined practical aspects of his field with theoretical insights. A steadfast advocate of the Hydraulics Laboratory, he promoted its cause at every turn and seemed never to cease his push for its expansion and betterment. He expressed concern for numerous aspects of IIHR's functioning, from his students' morale and character, to the use of tools, to the Hydraulics Laboratory's physical appearance.

His meticulous care as well as his high standards are revealed by a glance through his administrative papers. He carefully recorded, for example, the "notable visitors" who entered through the Hydraulics Laboratory's portals each year (although he made no comments on how he had divided the notables from the unnotables). Tallying the results, he surmised that a total of 184 notables had visited in a decade, with Iowa and Illinois by far outnumbering other source locales. He delivered written instructions to students on report writing, counseling them that "the preparation of an orderly and well worded report . . . is the best possible training for the student in the principles of engineering report writing," and warned them to prepare reports ". . . not as a tedious clerical exercise, but with neatness and pride, as though the manuscript were being prepared for publication or . . . the information was worth hundreds of dollars."

His concern extended to the details of the laboratory's operation. He stipulated that no experimental work should be conducted on the Sabbath, when the laboratory would be duly closed. Memoranda advised librarians in the cataloguing of books: he explained that he was attempting "to make recommendations which would restore order out of chaos." He detailed the construction of an educational hydraulics exhibit for Iowa's state fair. It seemed that everything having to do with the Hydraulics Laboratory was to be performed with consideration of its effects on the research, the researchers, the laboratory itself, and the larger community, as if all were tied to the highest expression of the human experience.

Nagler seemed to carry out his numerous and multifaceted professional activities without loss of commitment to church, family, and community. He never stopped going, acting, doing. Active as a church officer and Sunday School superintendent, he also counseled his church's youth, worked with the Boy Scouts, and became an advocate for the newly formed state-park system. He tramped the site of the future Lake Macbride State Park just north of Iowa City, and here as well as on other state park sites, he assessed the potential for artificial lakes and then designed the dams that would form the lakes.

He married in 1921 and fathered three children. He designed and built for his new wife a house near the Hydraulics Lab, on Melrose Circle, which was later occupied for a year by IIHR director Hunter Rouse. To fit his work and other outside activities into his home life, Nagler frequently took members of his family along wherever he went—to professional meetings in New Orleans and New York, to potential Iowa dam sites near Spillville, Anamosa, and Pikes Peak State Park. Nagler built his tiny son Robert a seat on the push-type lawn mower so that father could explain matters to his son while mowing the grass—things about the clouds or the weather, for example. By age eight, Robert was trudging off to the Hydraulics Laboratory with his father, where the two of them "got their hands wet" in the experimental channel.

Nagler was a large, powerful man who routinely returned from his ambles in the field loaded down with the rocks that he had collected. He used these to build a large rock garden with a fountain, a waterfall, and three interconnected pools. But his unusually good health, strength, and vigor could not prevent a normal attack of appendicitis, which was diagnosed at first as the flu. After disregarding his supposedly minor illness for a few days, Nagler checked into the University Hospital a few blocks from his home and underwent emergency surgery, but by that time his appendix had ruptured and the infection had spread. While his premature death might have been prevented with sulfa drugs or modern antibiotics, those were

Nagler would fit his three children into the back seat of his Dodge touring sedan and take them along on his field investigations of Iowa's rivers and potential dam sites. He'd return to town with his car loaded with boulders that he had collected for his rock garden.

not available in 1933. He survived a second surgery two weeks later, when portions of his liver were removed. Nagler died a few weeks after that on November 10, 1933. His grave in Oakland Cemetery is marked by a massive uncut granite boulder, engraved simply with his name and the years of his life, selected by his wife to characterize his love of nature and rocks.

3. Research and the Establishment of IIHR, 1920–33

During Nagler's tenure, the field of hydraulics rapidly expanded in importance and breadth of activity. Hydraulics-related problems were abundant and obvious. As Nagler himself wrote, "The field open to hydraulic research is almost unlimited." Laboratories were popping up across the country at major universities, encouraging the assemblage of collaborating researchers and enabling the evolving use of small-scale models, which in turn stimulated an attack on basic problems that had not previously been investigated. Spin-off projects were attacked with zest, for hydraulic research was providing useful and necessary answers to basic questions and thus was well funded. All these factors combined to feed one another, the funding permitting the expansion of laboratories and personnel, which jointly fostered the more sophisticated use of models, resulting in a recognition of additional research problems that beckoned the researchers ever onward. Even the most straightforward problems commanded a sense of excitement and energy.

Research during this period focused on questions that until then had remained largely unaddressed. How should an outlet of a pipe be shaped? How do curves in a river channel affect the velocity, scour, and pressure of the water? And what traits typify water flowing over stream obstructions (called weirs)? During Nagler's tenure, IIHR attacked all these questions, typically under contracts from a government agency, private company, or engineering firm that desired their applied results (although IIHR graduate students remained engaged in more theoretical challenges or, as Nagler explained, ". . . in the study of more academic problems"). Nagler and his colleagues examined the water-carrying capacity of various cul-

Research in the 1920s was often directed toward answering basic questions about practical problems, such as how water could most efficiently be transported through culverts of various shapes and materials. Full-scale culverts such as this 3-by-3-foot, end-flared, concrete box were constructed in the channel to the north of the laboratory, the water gates opened, and flows through various culverts compared. The results were fed directly into plans for highway improvements.

vert shapes and materials. They tested the performance of meters that measured current, assessed characteristics of the hydraulic jump (the transition from shallow, fast moving water to deep, slow moving water), and investigated how metal flap gates affect water flow. They observed how bridge piers and abutments of various shapes block or constrict water, causing local scouring of the riverbed. Sometimes the most basic questions revealed surprising results. For example, Nagler's team discovered that in a given time period, almost twice as much water could pass through a smooth concrete culvert as through a corrugated metal culvert with the same inner diameter. Or (as Nagler pronounced with pride), the 57 unlikely solutions that had been proposed to abate flood problems at Milan, Illinois, had been combined at the Hydraulics Laboratory into a single successful resolution.

The 1922 addition of David Yarnell to the research team steered

IIHR's research in new directions. Officially, Yarnell was an engineer with the U.S. Department of Agriculture who was stationed at the Hydraulics Laboratory. He and Nagler collaborated on culvert, river-bend, bridge-pier, and other research for many years. Several years later, collaborative research with other government agencies was promoted by the fact that these agencies rented office space in the Hydraulics Laboratory and thus had ready access to lab personnel as well as equipment.

Then, in 1928, Nagler was pulled into working for the U.S. Army Corps of Engineers (COE) on comprehensive surveys of the Mississippi River's tributaries. The Hydraulics Laboratory had already undertaken Mississippi River work in 1924, when it had measured the flow at the Keokuk dam and modeled the dam's spillways for the Mississippi River Power Company. The COE's surveys addressed questions of flood control and commercial use of the Mississippi's drainage basin in Iowa. With the studies of Ralston Creek, which had commenced earlier, and the assessments of Iowa rivers performed for Iowa's Board of Conservation and Fish and Game Commission in later years, river surveys comprised a major emphasis at the laboratory and the single greatest focus of Nagler's career.

Nagler's Mississippi River work also carried both IIHR and the COE—as well as the Mississippi itself—in major new directions. Nagler, on leave from the university, spent many hours in the COE's Rock Island offices preparing his survey reports. While there, he entered discussions about new plans to create cheap and reliable Mississippi River transportation. In previous decades, the COE had attempted to establish first a 4.5-foot-deep channel, then a 6-foot channel, by constricting the river into a narrower (and supposedly ever deepening) course. These efforts had failed to provide the hoped-for dependable navigation channel. Once loggers depleted the Northwoods forests in the early 1900s and rafts of white pine logs stopped floating downstream to lumber mills, through traffic on the Mississippi dried up and disappeared.

By the 1920s, private sources were rallying for restored commerce on the Mississippi. Large-scale transportation would be possible only if the river were to be deepened, and that would require the construction of dams to hold water at the required depth and locks to allow the passage of boats and barges. Nagler, while work-

In 1927, when this photograph was taken, IIHR's largely industry-supported, applied efforts exceeded the capacity of the original, minuscule Hydraulics Laboratory. Hydraulic equipment and research activities overflowed into this testing laboratory, which was located next to the Engineering Building on the east side of the Iowa River. Shown here is the weight tank, which is still in use.

ing on his river reports at the COE offices in Keokuk, was drawn into discussions about how to create the deep, reliable channel. He reportedly was one of the first to recommend the channelization program that was later carried out, and he spent considerable time on computations related to the creation of a 9-foot channel. Nagler also advised the COE on where to place its dams. Congress approved the 9-foot channel in the early 1930s, funding the construction of numerous locks and dams shortly thereafter.

Nagler suggested the use of models of these Mississippi River locks and dams, and he helped start the COE on the model-testing program that remained in place through the twentieth century. The dam at Hastings, Minnesota, was already under construction in 1930 when the COE recognized the benefits of testing the performance of those sections being built on a sand bed. Government engineers were sent to Iowa City to work with Hydraulics Laboratory staff, who modeled the dam along with a 20-mile stretch of the river.

The results were sufficiently productive to establish the Hydraulics Laboratory as a major modeling site for federal projects in the eastern United States. Soon models for Upper Mississippi River locks, dams, or spillways near Red Wing, Onalaska, Alton, Canton, Trempealeau, and elsewhere, as well as models for structures on the Ohio, Tennessee, Monongahela, and Illinois Rivers, speckled the Hydraulics Laboratory.

Model testing for the COE's Mississippi River locks and dams would become one of the Hydraulics Laboratory's major focuses in the 1930s, bringing in a substantial portion of its consulting revenue. Thus, Floyd Nagler and IIHR, in its early years, were instrumental in the massive restructuring of the Upper Mississippi that continues to characterize the river and its use.

4.   Building the
       Hydraulics Laboratory

Hydraulics is among the oldest of sciences. Practical knowledge of moving water was prerequisite to the correct functioning of irrigation systems, which in turn constituted a major network tying together the ancient, complex civilizations of Mesopotamia 6,000 years ago. And if a hydraulics laboratory is defined as any site where practical applications of water in motion are studied, such laboratories too have an ancient history.

However, the advent of the modern hydraulics laboratory dates only to approximately a hundred years ago, at least in this country. Although water provided the power, transportation, and sustenance that allowed the nation's development, water movement was little studied under controlled conditions until a flurry of labs was constructed in the late nineteenth and early twentieth centuries. Their appearance was stimulated by a multitude of factors. Foremost among these was society's growing recognition of the need to understand water flow. Ever larger, faster, and more numerous ships were plying the earth's waters. The spread of electrification required the construction of hydroelectric plants with larger dams and reservoirs. Growing cities, consuming ever more water, needed clean water sources and sites for sewage disposal. Providing adequate water for the soaring populations of the dry western United States was especially challenging. And vulnerable floodplains everywhere were being clogged with ever larger industrial structures and population centers that demanded protection from the rivers that provided their water and transportation. Such types of development required predictive laboratory studies of rivers and water flow. The traditional trial-and-error approach for undertaking local engineering works simply was not adequate for the increasingly sophisticated projects contemplated by society.

In addition, laboratory studies were furthered by a growing understanding of fluid mechanics and by a recognition of the benefits of small-scale, proportional laboratory models for testing real-life processes that were too large or too complex to be worked with in the field. By 1922, over four dozen hydraulics laboratories with experimental facilities dotted the American landscape, each reflecting regional needs and pride; all had been established within the previous four decades. Three-fourths of these were located at educational institutions. The others were operated by the government or by private companies, the latter often for the purpose of testing and developing turbines, ships, or other machinery. Most labs at educational institutions existed to allow students to practice the use of measuring instruments. Only a handful performed basic research, and Iowa was one of these few.

The construction of the University of Iowa's Hydraulics Laboratory fitted into the early-twentieth-century proliferation of laboratories and the nation's recognition of the need for hydraulic research. Its inception dates back to Iowa City's earliest days, when in 1843 Walter Terrell built a three-story gristmill and low dam (the Iowa River's first) about a half mile north of town, across the road from today's Mayflower Residence Hall on North Dubuque St. Early settlers, dependent on locally processed goods, came from distances of 50 to 75 miles to have their grains ground into flour and their wool carded for spinning, until a flood damaged the dam in 1881 and the mill subsequently went bankrupt. Finally, in 1903, the remainder of the dam was ripped from the river and the water rights were donated by Terrell's daughter to the university.

As early as 1906, the university had anticipated the creation of a hydraulics laboratory on campus. That year, the university built a dam across the Iowa River just south of Burlington Street and slapped a hydroelectric power plant at the east end. At the west end of the dam, a 10-foot-wide gap was retained through which water could enter a future experimental channel. When the city constructed the Burlington Street concrete arch bridge over the river in 1914–15, the newly created Department of Mechanics and Hydraulics was asked to design a retaining wall extending south from the west end of the bridge. The department included in its design a 130-

The original Hydraulics Laboratory, constructed in 1919. When the Iowa River was dammed in 1906, a slot had been left in the dam's far end to feed water into a hydraulic research channel.

foot-long concrete experimental channel, 10 feet wide and 10 feet deep, along the wall's base. The channel was erected largely by engineering students in 1918–19, along with a squat, square, brick workshop, a mere 22 feet per side, perched atop a concrete pedestal over the open channel flume. This hydraulics laboratory demanded a director. Floyd Nagler answered the call and assumed leadership in 1920.

Iowa's Hydraulics Laboratory initially was conceived as a turbine-testing laboratory, but this notion was rapidly displaced by investigations of numerous types of flow instruments, hydraulic structures, and water-flow processes. Studies performed within the lab capitalized on the large experimental channel. While the lab may have been small, the channel was of a "remarkably generous scale," a feature about which Nagler continually boasted. Because the experimental channel drew its water directly from the Iowa River and a hefty 3,140-square-mile upstream area, the lab had access to a robust gravity-fed water supply under moderate head and could perform experiments on a larger scale than most other labs.

The channel was accessible either from outside the building or through the lab's removable wooden-plank floor.

From the very beginning, the Hydraulics Laboratory beaconed success. Demand kept the lab continuously engaged. Local newspapers regularly commended it for the quality as well as quantity of its efforts, citing it as the best equipped hydraulics lab in the country. Nagler himself frequently went on the stump for the laboratory, writing or speaking to promote recognition of its qualities. Referring to his modest 484-square-foot workshop, he bragged that "... there are but few experiments in the history of hydraulics that could not be duplicated in this laboratory and many others that can be executed at this place more readily than at any other laboratory in this country." All the while, as experiments and consulting flourished, students gained valuable hands-on experience with equipment and real-life projects. The lab provided both abundant demonstrations for undergraduates and research possibilities for graduate students and their professors.

Soon the tiny workshop did not allow adequate room for experiments. By 1923, reports of lab activities bemoaned the fact that the lab was forced to reject proposals because of its crowded schedule, and additional laboratory space for teaching and research was sought across the river in a structure near the Engineering Building. Nagler, a man of unlimited energy, was not one to allow limitations to go unchallenged. If his lab's work required more space, he would get it.

And get it he did. The year 1928 marked the opening of a new 60-by-30-foot laboratory, intended as the initial wing of a larger future building. With the lab now three stories plus a basement, experimental apparatus occupied all but the uppermost level (which held offices and classrooms) and included flumes, weighing tanks, measuring basins, and a pump room and circulating water system to supply water to all floors as needed. With such equipment, small-scale model testing could proceed on an expanded scale. Large-scale tests continued in the channel, now extended to 190 feet in length, through which Iowa River water still passed underneath the building.

However, despite the Hydraulics Laboratory's growing renown, Nagler did not remain satisfied for long. His energetic and consis-

A growing research program soon led to space problems and to the planning of an expanded laboratory. Shown here is the north wing, constructed in 1928. The Hydraulics Laboratory's central tower and south wing were added four years later.

tent drive to promote the lab meant that the grants and contracts kept coming, making even the larger lab inadequate for the tasks at hand. In early 1930, he wrote:

> Throughout the year, it has been necessary to carefully schedule all experimental projects in order to avoid interference, and on some days, the laboratory has been operated continuously for 24 hours when graduate students and others have been forced to perform their investigations during night hours because of lack of space and water during the day time.

By the end of that same year, he was sending the engineering college's dean architectural sketches for an enlarged hydraulics lab, one with more room for offices and classrooms as well as larger equipment. He prefaced the sketches with a lengthy memorandum supporting the enlargement, arguing for it on the basis of the current lab's relatively minuscule size, the large and increasing number of graduate students (15 in 1930–31), and the project funds gleaned from industries, private engineering firms, and government agencies (primarily the U.S. Army Corps of Engineers and the U.S. Department of Agriculture). These projects, he was quick to point out, cost the university a mere $2,726.85 during the previous decade, despite the many benefits they brought in terms of research opportunities for staff and students and the more than $100,000 in payments they fed into university coffers.

Nagler's persuasive charms must have mingled with the tenor of the times, for suddenly things started happening with amazing rapidity. In the summer of 1931, a 4-foot-diameter pipe was installed to deliver water from the dam to a new, 16-foot-wide experimental channel under the lab. Toward the end of 1931, the Iowa Institute of Hydraulic Research was formally organized, complete with a six-man research staff, an equally large board of consultants consisting of faculty from throughout the university, and an advisory committee of engineers from outside locations. Formally transforming the laboratory into an institute increased the lab's stature and also simplified business operations. The newly formed IIHR remained an integral part of the university's College of Engineering, but it could independently negotiate its multiplying grants and contracts with other organizations and agencies. This arrangement holds to the present day, as do arrangements for IIHR's support. Project contributions allow IIHR to be largely self-supporting.

Finally, in 1932, a five-story central tower and a three-story south wing were added to the Hydraulics Laboratory's existing north wing, more than tripling the 1928 lab's floor space. Along with the other improvements, the 10-foot-wide channel was extended to 311 feet and the 16-foot-wide channel to 120 feet. This large new building was financed in part by federal agencies that rented space in the Hydraulics Laboratory. The U.S. Army Corps of Engineers had established a sub-office in the lab in 1929 to begin a modeling program on dams, spillways, and navigation structures. Following the final lab expansion in 1932, the U.S. Geological Survey and Weather Bureau also rented office space in the Hydraulics Laboratory.

Had Nagler not died prematurely in 1933, one senses that he might have continued expanding the Hydraulics Laboratory until it challenged even today's huge University Hospitals and Clinics complex in size. Even before the final expansion was completed, Nagler was protesting that the experimental projects "unfortunately . . . scattered in unoccupied buildings around Iowa City . . .

The next Hydraulics Laboratory expansion? Had Floyd Nagler not died prematurely, he may have continued his building expansions indefinitely. He joked that he would do so with this simulation, which he fabricated the year of his death by cutting and pasting together two actual photographs of the laboratory. He labeled the composite, "Hydraulics Laboratory as it will appear after the next addition."

are more than sufficient to fill the new addition, . . . [which] will soon be as cramped for room as the present [building]." His pride in IIHR remained unfailing to the end. "At the present time no American university has a laboratory with better facilities than Iowa's. When the [1932] addition is completed, the laboratory will probably be unsurpassed anywhere."

But as it was, Nagler's death a scant year later meant that his final expansion was also the Hydraulics Laboratory's last, and that for the rest of the century the outer IIHR building would look much as it did at the time of its completion. However, looks seldom reveal content, and just as rivers flow ever onward, so IIHR has continued its evolution. While the outer Hydraulics Laboratory stood firm as a substantial brick-and-mortar structure fused to a bedrock base, Nagler had designed the inner workings to be flexible, with large open experimental spaces in the basement, first, and second floors free of fixed apparatus, so that equipment could be modified as research projects evolved. A model of a dam spillway might yield to a model of a metropolitan storm-sewer drop shaft; a ship lock might be replaced by a fish-bypass system. Through the years, more permanent renovations have also become commonplace. In the 1950s, for example, the 311-foot open channel that Nagler had once praised for the volume of its flow was covered and converted into a ship-model towing tank.

While the outer integrity of this now-historic structure is maintained and has become respected as an icon of hydraulic research around the world, struggles continue to adapt its limited space to serve the needs of an ever-growing staff. The Hydraulics Laboratory still serves as IIHR's administrative center, with models and experimental equipment of past years having largely given way to offices, the Nagler Reference Room, and computer laboratories. Efforts to provide adequate and safe power through rewiring the building have proven to be major undertakings primarily because of Nagler's thoroughness and insistence on quality. While he had the vision to design a building so enormously sturdy that all floors could simultaneously support experimental equipment fully filled with water, he could not have foreseen the difficulty in drilling through several feet of concrete and steel I-beams to modernize electronics and communications networks. Meanwhile, IIHR's laboratory space has been consistently increased through the construction of annexes: the West (earlier Model) Annex (used from 1948 to 1986), East Annex (opened in 1975), Oakdale I Annex (1979), Model Annex (1982), Wind Tunnel Annex (1984), and Oakdale II Annex (1996). These structures are located in the map following the preface. In these structures, physical models continue to be constructed, tested, torn down, and rebuilt time and time again, making the contents of these laboratories as fluid as the liquid processes they study.

5. Tracing Iowa's Water Flow

Floyd Nagler seemed to have an absorbing love of free-flowing water and a fascination for its transformation into "usable" form through human constructs. Although he studied water flow in pipes, through curving laboratory flumes, and around indoor models of bridge piers, his papers consistently reflect a preoccupation with the unconstrained flow of water. Nagler's publication list is sprinkled with assessments of the passage of water through Iowa, including climatologic studies. "The Drouth of 1930 in Iowa," "The Water Yield from Small Watersheds in Iowa," and "A Survey of Iowa Floods" are but a few examples. He continually advocated statewide funding of "an adequate stream flow investigation program." Writing to Iowa's governor that failing to fund such a program would be "a very unfortunate matter—almost a calamity—for the future development of the state," he managed to reinstate a state-sponsored stream-gaging program that had been cut from the budget several years earlier. Nagler firmly believed that knowledge of our streams was imperative for harnessing the benefits of our waterways and decreasing their damage. "The information obtained from the adequate measurement of the flow of streams is almost indispensable in the economic design of a large variety of projects that directly or indirectly affect the public welfare, such as drainage, river polution [*sic*] by sewage, flood protection devices, waterways, . . ." he said in his letter to the governor.

Nagler's interest in the flow of water through Iowa and its effects is best exemplified by his river surveys, which commenced in 1922 with Ralston Creek. This creek drains approximately three square miles on the eastern edge of Iowa City. Although later encased in subdivisions, in Nagler's time Ralston Creek gurgled through farms outside the city limits. At that time, no data on entire small Midwestern drainages could be found. Nagler thus initiated a comprehensive study of this watershed, which he believed was representative of all such small drainages. Obtaining funding initially from the U.S. Department of Agriculture and the U.S. Geological Survey, he commenced regular measurements of rainfall, water flow, and groundwater, and surveyed the watershed's topography, soil, and crops. He trusted that the Ralston Creek watershed would reveal how the total stream system functioned, including "the intensity of midwestern storms, . . . concentration of flood runoff, . . . influence of vegetation on stream yield, and rate of infiltration of rainwater" among other features.

The lengthy sheets of hand-recorded data testify to Nagler's farsighted vision. At that time, when researchers were barely beginning to understand the complexities of integrated systems, Nagler had the foresight to examine many of the interconnected paths of the hydrologic cycle, and to do so for an extended time period. While modern computer-aided distributed watershed modeling studies enable examination of watersheds in great detail, Nagler was perhaps thinking of something even more sophisticated: deriving an understanding of individual responses throughout all component parts of the watershed. The Ralston Creek project was certainly one of the first studies, if not the very first, to attempt to do so. These data, collected until 1988, also provide one of the longest highly detailed records we have of the workings of a watershed. Ralston Creek data were used in 1933 to support the conclusions of a landmark paper on water infiltration, and since then have been used to verify computer models, as a foundation for theses examining one aspect of a watershed, and as the basis for several other journal articles. They remain a little-tapped gold mine of information on changes in the hydrologic regime that occur with urbanization and with altered farming practices—topics of major environmental interest in the 1990s.

The Ralston Creek studies under way, Nagler set out to investigate larger rivers. Just 50 miles to the east of the Hydraulics Laboratory pulsed the mighty Mississippi, part of Earth's third-longest river system, which drains 40% of the nation. Seasonal surges and their resulting floods had always plagued riverside settlements.

Now the Rock Island District of the U.S. Army Corps of Engineers (COE), the agency that for decades had attempted to train this mighty flow, was asked to report to Congress on the best method for controlling these floods regionally—and, incidentally, on possibilities for navigation, power production, and irrigation as well.

In 1924, Nagler had measured the river's flow at the Keokuk Dam spillways. Now he was asked to provide expert aid to this new project. He asserted that the Mississippi could best be understood by learning more about its tributaries. In May 1928, he became an enthusiastic "Engineer in Charge of Stream Investigations" for the COE and initiated comprehensive surveys of Iowa's rivers and streams. Two field parties under his supervision helped him assess stream profiles and features, report on prospective reservoir sites, and search out all power developments. A third crew provided office support back at Rock Island.

Although Nagler took a leave of absence from the university that summer to work on the river surveys, he energetically continued to supervise the Hydraulics Laboratory. He did the same again from May 1929 to the end of 1930. After spending the summers in the field, he worked 10 to 12 hours each winter day spinning out detailed, lengthy reports (each typically well over 100 pages) on the Iowa, Des Moines, Boone, Raccoon, Turkey, Wapsipinicon, and nine other rivers. In these reports, he comprehensively examined flow characteristics and sediment, floods, and control measures for each river and its tributaries, as well as climate, land use, and population for each watershed, backing up his words with numerous charts, maps, and photos. He located, described, and estimated the effects of structures along and in each river, including levees, dams, water-power developments, and river-training projects. And he detailed his conclusions about the desirability of promoting navigation, irrigation, flood control structures, and power developments.

Nagler's work on the river-survey project ended in December 1930. By that time, Nagler had laid the groundwork that established IIHR as the major site for modeling locks and dams on the upper Mississippi and other rivers, and which would link IIHR and COE efforts through the coming decades.

With his efforts for the COE completed, Nagler did not remain unoccupied for long. His surveys had deepened his intimate knowl-

This 1924 study of water flow just upstream of the Keokuk Dam was the first of IIHR's many efforts focusing on the Mississippi River. The elaborate metal framework above the researchers, designed specifically for this project, dropped the current meter into the rapidly flowing water and managed to obtain accurate readings under difficult conditions.

edge of the streams of Iowa, and that knowledge was much in demand. In 1929, he had been called upon by the State Board of Conservation for advice on proposals to dam or otherwise modify certain rivers. In 1932, he started consulting with Iowa's State Board of Conservation and the State Fish and Game Commission about a variety of water-related projects.

While Nagler's COE-sponsored studies had sprung from the mandate to develop the nation's waters, surveys for the two state agencies were conceived with the opposite goal in mind: that of conserving those natural resources that remained. The early twentieth century was marked by a ground-based conservation movement. People throughout the nation had come to realize that our country's settlement had proceeded largely through the despoiling of natural resources that now, where they survived, needed to be preserved. A number of state initiatives resulted from this movement, one of these being a state-funded comprehensive survey of Iowa's natural resources leading to a long-term conservation plan. The 1933

Twenty-Five Year Conservation Plan for Iowa, a massive effort that re-organized the state's conservation priorities and agencies, was the result. For this report, Nagler served as consultant for hydraulics and hydrology, explaining fundamental considerations regarding artificial and natural lakes and the construction of dams for power projects. He inspected the sites of dozens of potential artificial lakes in future state parks, then wrote reports on the feasibility, aesthetics, and estimated costs of constructing those lakes. And he inspected and made recommendations concerning management and "improvement" of an equal number of natural lakes.

When the *Twenty-Five Year Conservation Plan* was completed, he continued similar consultations with Roosevelt's Civilian Conservation Corps, inspecting and reporting on sites where the CCC was considering constructing or improving an Iowa lake or dam, including the dam that would form Lake Macbride just north of Iowa City. A number of these projects were under construction at the time of his death.

One senses in his many river and lake reports that Nagler dreamed of an Iowa speckled with waters both stopped and flowing, an Iowa jeweled with lakes and dams where any free-flowing water was put to service for the betterment of Iowa's peoples. An engineer at heart and in practice, Nagler seemed intent on harnessing the broad potential of Iowa's water to provide the basic human needs to the maximum extent possible. Perhaps his greatest passion here was using water to provide power. His files abound with reports prepared for various power companies regarding the operation of their water turbines and the construction of new hydropower plants. Whenever surveying an Iowa stream, he assessed its suitability for providing hydroelectric power. While his stream studies for the COE firmly stated that he saw little desirability in using Iowa's rivers for irrigation or commercial navigation, he thought differently about waterpower developments. For the Iowa River, for example, he recommended increasing the output of existing plants and designated sites for 24 new power plants (although he admitted that only four were really feasible). His bias toward the development of waterpower was revealed in a paper entitled "Will Water Waste or Work," in which he wrote of resource-

Nagler exercised his fascination for waterpower by promoting hydroelectric plants in Iowa and by collecting and exhibiting turbines from abandoned mills. When soliciting these discarded turbines from a power plant in Greene, Iowa, he wrote to its owner, "I wonder how much persuasion it would take to get you to load these three wheels, complete with their guide cases, onto a Rock Island train and donate them to the hydraulics laboratory at the State University of Iowa. They would certainly be of interest and value to my students. . . . When I finished my undergraduate work, I had no conception of what a water wheel looked like, and I am determined that our students will know water wheels from 'a to z.'"

depletion concerns remarkably similar to those of the late twentieth century:

> With only five percent of the waterpower of the world developed at the present time while the world's coal supply is being depleted at the rate which will mean its entire exhaustion in 2,000 years, the question of whether water shall work or waste presents a real challenge to the engineer-economist. In Iowa alone, and Iowa has comparatively little waterpower, there are 200,000 horsepower which can yet be developed.

In a report published the following year, he had increased his prediction of expected Iowa horsepower to 400,000, and wrote that this waterpower would "play an important part in the industrial development of the state" and save more than two million tons of coal per year. Here and elsewhere he admitted that the harnessing of such waterpower in Iowa would not be an easy task or cheap; yet he maintained that waterpower plants would be economical when tied into large electrical systems also fed by other power sources, and he pinpointed specific sites for dam construction and development of power plants.

Nagler's fascination with waterpower also led him backwards into history (see chapter 10). He loved to wander the Iowa countryside, picking up a rock here or there, and searching out remains of earlier times. In particular, the old water mills grabbed his fancy. Nagler himself rooted out old millstones from the mud and sought out rusting turbines in scrapyards, arranging for these signs of earlier times to be transported back to Iowa City, where he proudly displayed them. His river surveys allowed him ample opportunity to exercise these passions; his work with the COE provided him with two field teams that he instructed to do likewise. These teams were directed to locate and photograph as many old water mills as they could. Nagler then put the slides together into a lecture that he presented repeatedly to the general public. In this way, Nagler's river surveys allowed him to express a passion for historical studies that was to become an IIHR tradition and has been reflected by IIHR directors ever since.

IIHR staff, 1945

II THE WIDENING STREAM
The Middle Years
1933–1966

6. The Depression, Hunter Rouse, and Fundamental Fluid Mechanics

In June 1985, IIHR director John F. Kennedy wrote a congratulatory letter to Hunter Rouse for one of his many honors. Kennedy signed the letter as "the man who is still introduced (proudly) as the Director of Rouse's lab."

This pairing of place with person occurred between 1944 and 1966, when Hunter Rouse, IIHR director, tacked hydraulics firmly to its scientific and mathematical underpinnings by tying it in multiple ways to the study of fluid mechanics. His writings, research, and ideas permeated the professional community so broadly that Rouse, rising to prominence as a preeminent hydraulician, pulled the entire institute into the international spotlight. The prominence that Rouse attracted to himself, his ideas, and his laboratory was sufficiently intense that his followers tended to forget about the years that came before and the researchers who preceded him.

Yet despite Rouse's unquestionable significance, IIHR was founded and fostered by the labor and dedication of many others. Some of these forgotten heroes are the leaders and researchers who nursed IIHR through the 1930s, when Floyd Nagler's death left the institute in shock and when shadows of the Great Depression spread across the nation, casting doubt upon the viability of this institution and many others. The institute—a mere fledgling at the time—hobbled from one project to another, for a time piecing together leadership and programs, until new defense-related research initiatives and funding were sparked by the flames of World War II. At that time, fundamentals were thrust to the forefront of the institute's agenda and were held there by the serendipitous juncture of Hunter Rouse's dogged determination and the nation's needs and desires. Basic research in the physics of fluid flows soared in quantity, stature, and funding, and fluid mechanics was inaugurated into the curriculum, joining classes in applied hydraulics. All in all, the period between 1933 and 1966 was one of turmoil and transition, famine and feast, when economic and political events outside the institute's walls were clearly reflected on the waters of the flumes and tanks inside its walls. This was a time of extremes: when a small institute became large, when meager research funding swelled to plenty, when the focus in teaching and research swung from the practical and applied to the theoretical and basic, and when leaders dedicated to the multiple practical issues of the day were replaced by one man who single-mindedly created a vision of merging hydraulics with fluid mechanics and assumed total responsibility for its execution.

Hunter Rouse's hearty desire to bridge the gap between physics and engineering and between theory and experiment is a major theme that will permeate this and following chapters, just as it permeated the institute a half century ago. But before this topic is discussed in more detail, we must return to the decade immediately following Floyd Nagler's death, examine the tenor of the times, and memorialize his immediate successors.

Floyd Nagler, a tower of strength and energy who was never ill, was a mere 41 years old when he succumbed to a ruptured appendix in 1933. His death dealt a blow to the hydraulics profession nationwide. As one of his many obituaries stated, "He has put hydraulic engineering on the top shelf in the United States." His unexpected death must also have thrown the institute into chaos. Nagler had single-handedly nurtured the Hydraulics Laboratory from a one-man show to a widely recognized operation that claimed prestige and a rising reputation. Nagler's colleagues naturally would have assumed that this visionary founder of IIHR would continue to guide his creation for decades to come. Thus, there was no natural person to step into his shoes, no one who had been groomed for the job, not even someone who had been at the lab long enough to have matured professionally within its walls. In fact, the number of academicians at the institute still remained rather small. Professors Joe Howe and Chesley Posey had recently joined the institute's research staff and would remain to become some of the longest-term IIHR faculty members; their arrival had been preceded by a few years by Professor Mavis, who with Howe, Posey, and Nagler (and their successors) also was a faculty member in the Engineering College's Department of Mechanics and Hydraulics. Additional institute offices were occupied by nonacademic engineers who represented specific governmental agencies and worked here on special hydraulics-related projects. While they were about three times as numerous as the university employees, these engineers were not responsible for the laboratory's management. Thus, IIHR had relied heavily on Nagler and his vision and drive. Nagler had spearheaded the majority of institute initiatives, and his departure would end new innovative enterprises for several years.

The only obvious replacement for Nagler seemed to be Sherman Woodward, first chair of the Department of Mechanics and Hydraulics, who had hired Nagler and consistently supported the Hydraulics Laboratory. Woodward assumed the reins of control for a brief eight months, after which the institute was orphaned a second time. Woodward, in addition to holding his professorship, had been the head of a local bank that failed during the Great Depression. The failure financially wiped out Woodward, and in July 1934, he moved on to the Tennessee Valley Authority (TVA), there bringing in a much higher salary and attempting to recoup his losses.

A new approach was then attempted for the institute. Administration of the lab was split between the dean of the College of Engineering and an institute-based faculty member, guided (as before) by a group of advisory consultants. Thus, IIHR, which had always been an integral and significant part of the university's College of Engineering, became even more so for a decade, until Rouse became director in 1944.

This split administration—so unlike Nagler's one-man directorship—may have in a sense required no one person to take strong charge of IIHR and forge a new course. The depression also must have taken its toll, eating into funds and forcing all to search for financial security where they might. Whatever the reason, whether by force of external circumstance or because of the personalities of IIHR's administrators, in this difficult decade the institute seems to have been more reactive than proactive, more seeking to hold its own than to forge forward to new lands. Floyd Nagler had constructed the Hydraulics Laboratory and established IIHR. He had

Sherman Woodward, IIHR Director after Floyd Nagler's death (November 1933) until July 1934.

envisioned a premier center for hydraulic research and had set the ship on its course. His immediate followers competently and steadily followed that course, educating students, nurturing the relationship with the U.S. Army Corps of Engineers, and welcoming similar cooperative agreements and applied contracts that carried the institute through the depression. A few new research topics were adopted during this period, but for the most part the 1930s seem to have emphasized maintenance rather than growth, repairing the old rather than creating the new, a tidewater decade sandwiched between the reigns of two dynamic directors who, in contrast, seemed to forge ahead into unknown territory with every breath they took.

The institute's split leadership consisted first of Dean Williams as director with F. T. Mavis as associate director in charge (roughly 1934–36), then Dean Dawson as director with Emory Lane as associate director in charge (roughly 1936–42), and finally Dean Dawson as director with Anton Kalinske and Hunter Rouse as joint associate directors in charge (1942–44). After this time, the administration resumed its previous structure, with Hunter Rouse assuming institute directorship in 1944 and taking full charge of institute activities himself. Rouse maintained that post until he assumed the college's deanship in 1966 and John F. Kennedy came to Iowa to direct IIHR.

Far less is known about the Williams-Mavis and Dawson-Lane duos than about their dynamic predecessor and successor. Their activities at IIHR remain for the most part gray and indistinct, their hazy outlines sharpened only by a few remaining documents that lend clarity. Dean Clement Clarence Williams had come to Iowa City from the University of Illinois in 1926, at age 44, to assume leadership of the College of Engineering following the sudden death of Dean Raymond. He remained here until he accepted a call in 1935 to become president of Lehigh University in Pennsylvania. A civil engineer with experience in railroads, structural engineering, utilities, and engineering administration and education, he displayed no evident inclination toward hydraulics. One can assume that during his directorship from 1934 to 1935, he left matters mostly to Associate Director Mavis.

Frederic T. (Ted) Mavis joined the Mechanics and Hydraulics faculty and the institute staff in 1928, immediately after completing his graduate work and a one-year stint as one of the first Freeman Traveling Fellows studying hydraulic laboratory practice in Germany. Mavis assumed associate directorship of IIHR in 1934 and became acting head of the Department of Mechanics and Hydraulics a scant eight months later, when Woodward left. Mavis maintained that dual administrative role until 1936, when he gave up his associate directorship to devote full attention to chairing the

Kent Collection, University Archives

Left: Clement C. Williams, Dean of the College of Engineering and IIHR Director (1934–35); right, Frederic T. Mavis, Associate Director in charge of IIHR (1934–36).

Mechanics and Hydraulics department. However, he did not abandon his involvement in hydraulics or the institute. His publications on topics such as hydrology, sand permeability, and culvert and pipe studies continued. Mavis also published some of the institute's first papers on sediment transport, a then-new field that has continued to receive attention at IIHR.

Mavis sought new ways to improve the welfare of the institute and to expand knowledge of hydraulics. His efforts often seemed to be directed toward organizational matters. He successfully lobbied the university to purchase private land lying south and west of the Hydraulics Lab. This land, he claimed, not only was necessary for parking space, it also could provide space for an outdoor laboratory for river models and rating current meters and would prevent the construction of a Super Service Oil Company filling station, which a Mr. Finnigan was threatening to build there. (When his plan failed, the local builder in retribution petitioned the city government to tear down the Hydraulics Laboratory because it "encroaches on the public highway." His pleas obviously went unheeded.) Mavis also envisioned a mounted display of Nagler's turbine collection. He wanted these "museum pieces," then regrettably "little more than a rusty blot on the doorstep of the [hydraulics] building," artistically sprinkled about the land west of the Hydraulics Laboratory on random stone pedestals separating the highway and the parking area. Mavis's passion for obtaining parking space pervaded his writings: even when reporting the institute's impending doom from loss of crucial federal funding, he complained that the institute "does not even provide adequate space for parking the automobiles of regular employees"—as if this were the ultimate insult. Mavis's foresight assured that university-owned land was available for construction of the West Annex (earlier the Model Annex) in 1948. In 1938, Mavis arranged for Nagler's turbines to be painted and displayed on that site.

Mavis also composed an office manual explaining logistical details of running the institute. His 27-item "Regulations for the Use of the Hydraulics Laboratory" cautioned readers to turn off the lights, never to smoke in the building, not to waste power by pumping more water than needed, and not to dump refuse out the windows into the river. (This apparently common practice had "made the river front very unsightly.") In 1939, he edited the first in a series of bulletins describing the institute, *Two Decades of Hydraulics at the University of Iowa*. Perhaps most importantly, he conceived and organized the first national conference for hydraulic engineers, the Hydraulics Conference held at IIHR in 1939. Mavis left Iowa to head the Department of Civil Engineering at Pennsylvania State University soon after the conference that same year.

The hole created by Dean Williams's departure in 1935 was filled by Francis (Frank) Murray Dawson a year later. Dawson, a skilled educator and administrator, possessed the hydraulic expertise that Williams had lacked. Trained and experienced in hydraulics, he came to Iowa from the University of Wisconsin, where he had been professor of Hydraulics and Sanitary Engineering. He had co-authored a textbook entitled *Hydraulics* and arrived in Iowa as an established expert in plumbing research, with connections to industrial funders for that topic. He undoubtedly kept in close contact with the institute and its functions. The most obvious sign of this is IIHR's commencement in the late 1930s of plumbing research as a well funded research field, one that extended through and beyond the war and earned the institute the title of the official testing labo-

Left: Francis M. Dawson, Dean of the College of Engineering and IIHR Director (1936–44); right, Emory W. Lane, Associate Director in charge of IIHR (1936–42), Associate Director (1944–46).

ratory for the National Association of Master Plumbers. Dawson himself authored a number of reports on plumbing safety, water supply, and engineering education, and he continued experimental work at IIHR on plumbing through the 1940s despite his duties as dean. Dawson remained as a much loved and respected dean until he retired following a stroke in 1959.

Emory Wilson Lane, the former head of the Bureau of Reclamation Hydraulics Laboratory, came to the university as a professor of hydraulic engineering in 1935. Lane was brought to Iowa to fill Woodward's professorship. He accepted the associate directorship in 1936, when Mavis stepped down. An astute river and field engineer, Lane was an expert in sediment research and channel stability, and he greatly expanded the sediment studies that had been started by Mavis and would continue as an institute focus through the coming decades. His publications in the field were numerous. They also dealt with other river-related topics and with river matters in China (where he had worked), but his vision did not extend far beyond this purview. His future plans for the institute, as stated in his annual reports, were largely hopes that established research projects would continue to receive funding to carry IIHR through the depression. Although new fields were investigated during his tenure, these efforts were initiated by others: Dawson brought plumbing studies to the institute, fishways studies stemmed from Nagler's criticism of state policies several years earlier, and investigations of turbulence were pioneered by Kalinske.

In 1942, Lane requested a two-year leave of absence to follow Woodward to the TVA. The TVA at that time was busy building dams to power the refining of aluminum ore needed for the war, and Lane felt a patriotic calling to assist with that effort. Dean Dawson filled Lane's associate directorship temporarily with two rising stars who had come to Iowa in the late 1930s: Anton Kalinske and Hunter Rouse. Both were energetic in their explorations of various fields, attempting to lead the institute into new activities and areas of investigation. Both were visionary leaders whose assertiveness contrasted dramatically with Lane's relaxed approach.

Kalinske had studied under Dawson at Wisconsin. He followed Dawson to Iowa in the mid-1930s and then took a leading role in expanding Dawson's plumbing studies here. His publications on the subject were abundant. He also introduced himself to the literature on turbulence and commenced experiments to study the diffusion of dye in water through analysis of motion pictures. His efforts and prolific writings probed diverse other topics as well, such as sediment transport, turbulence, fluid friction, and multiple aspects of pipe studies, and provided focus and research material for numerous graduate students. When the war started, Kalinske recognized the opportunities for expansion of research funding. Rather than wait for federal agencies to come knocking, Kalinske took a summer job at the U.S. Navy's ship testing laboratory, the David Taylor Model Basin, established connections there, and started to solicit projects for the institute. In this way he initiated the institute's long-lasting investigations in ship hydrodynamics among other topics. Kalinske also instigated the construction of the institute's first research water tunnel in 1943, which set the stage for long-term, government-funded cavitation studies of projectiles.

Dean Dawson had brought Hunter Rouse to the institute in 1939. Born in 1906 in Toledo, Ohio, Rouse had received bachelor's and master's degrees in civil engineering at MIT and had studied in

Anton Kalinske, co-Associate Director (with Rouse) in charge of IIHR (1942–44), who was instrumental in stimulating turbulence and other fluid-mechanics studies at IIHR, and also attracted major new funding from the federal government.

Europe for two years. In 1932, he received his doctorate at the Technische Hochschule in Karlsruhe, Germany. He returned to work first as a hydraulics assistant at MIT, then as an instructor at Columbia, and finally as an assistant professor and Soil Conservation Service research engineer at the California Institute of Technology (CalTech). By the time he reached Iowa, his research and his first textbook had already won recognition. He came to Iowa as a professor of Fluid Mechanics and consultant to IIHR, with Dawson assuming that he would fill Mavis's vacated slot as chair of the Department of Mechanics and Hydraulics. But Rouse clearly preferred research to administration. Thus, from the start, under mutual agreement, Joe Howe performed the department administrative details for Rouse, an arrangement that was formalized when Howe became departmental chair in 1943. Rouse's energies were thus freed to reshape the institute's research program.

The maelstrom of war in Europe set whirlwinds swirling through the institute that, within a few brief years, would change the institute forever. From the start, Kalinske and Rouse stimulated interest in broader problems of fluid mechanics, particularly fluid turbulence, and broadened the purview of research projects beyond applied results into related fields of fluid motion. Their efforts attracted defense-related funding from the federal government, which expanded the institute's annual budget from a few thousand to tens of thousands of dollars. To facilitate the multiple war studies, graduate students were pulled from the classroom and enlisted as full-time assistants. New equipment (notably the institute's first wind and water tunnels) pulled fluid investigations into the study of flowing air. Numerous novel defense initiatives urgently sought methods for improving the Allies' ability to defeat the Axis powers, energizing and uniting institute staff as never before.

By the time Lane returned in 1944, presumably to resume the institute's leadership, the institute had grown well beyond Lane's abilities as a professor and sediment transport expert. Dawson recognized that Lane would have to be replaced. Professor Joe Howe later described the situation as extremely awkward. "Lane was a nice chap and very capable, but he just couldn't hold a candle to Rouse. . . . Rouse was clearly the best qualified. . . . Lane really wasn't doing a whole lot. With Rouse on the throne, we had all sorts of jobs for graduate students." If the funding for those jobs was to continue, Rouse would have to remain in control. To ease Lane's *de facto* demotion, Dawson ended the epoch of engineering deans nominally directing IIHR. He surrendered his directorship to Hunter Rouse, naming Lane once again as associate director. However, Dawson's intent was clear to Lane, who became extremely unhappy and began looking for other positions. In 1946, here returned to the Bureau of Reclamation, where he continued significant work on mechanisms of sediment transport.

Judging from his record, Kalinske, although less senior than Rouse, also had the makings of an excellent institute director. After nearly a decade of devotion to the institute, one can assume his disappointment at losing a position to match his contributions. Presumably dissatisfied, he likewise left in 1945 for a position in industry but many years later was again pulled into service to the university, when in 1967 Rouse, then dean, asked him to join the Engineering College Advisory Board as one of its founding members.

By the end of the war, Rouse was well recognized within the

Hunter Rouse, co-Associate Director (with Kalinske) in charge of IIHR (1942–44), IIHR Director (1944–65). Rouse (seated), knighted "the father of modern hydraulics" by his successor John Kennedy, exercised a single-minded passion for the fusion of fundamental fluid mechanics and classical hydraulics. An exacting teacher, he passed on his passion and high standards to his students.

hydraulics community, achieving prominence that 30 years later would induce Kennedy to quip, "Contrary to widely held belief, Hunter Rouse did not discover water, nor did he even invent hydraulics." To understand Rouse's larger-than-life stature and the significance of his contributions, we must stand back and examine hydraulics from a vantage point.

Hydraulics covers a variety of endeavors, ranging from practical applications such as river engineering and flood control to the abstract, highly mathematical, theoretical understanding of turbulent liquids, the resistance created as a fluid passes over a solid object, and so forth. Early in the century, in the United States, the discipline was almost totally dominated by one end of that spectrum: the practical. Floyd Nagler's research in river measurement and management was standard for the times, as were the classes taught in the Department of Mechanics and Hydraulics in those years: Water Power Engineering, Irrigation and Drainage, Hydraulic Turbines and Centrifugal Pumps, Flood Control, and the like. The practical demands of the problems at hand meant that the emphasis was usually on the experimental and empirical. Neither the available teaching materials nor the training practices of the times guided hydraulicians to alter course.

During this same period in Europe, however, the study of flow processes was starting to churn with new ideas. These were largely lumped into the rapidly emerging discipline of fluid mechanics—the physics of fluid behavior—that was evolving hand in hand with the development of human flight and its sister science, aeronautics. Modern fluid mechanics, with its origins in Ludwig Prandtl's 1904 boundary-layer paper and similar discourses, provided the tools required for an exacting analysis of flow over solid surfaces and around bodies such as airfoils and propellers.

Thus, for its first few decades, theoretical fluid mechanics developed in Europe primarily among mechanical engineers, ahead of the more immediate, practical developments in hydraulics that were the ongoing concerns of civil engineers. However, further advances in hydraulics required a better theoretical understanding of fluid flow. The juncture between the two was for a time delayed by the lack of hydraulicians with a strong background in theoretical fluid mechanics, and correspondingly by the paucity of fluid mechanicians exposed to the practical problems faced by hydraulicians. However, the merger between the two was inevitable. All that was needed was a bridge.

Hunter Rouse was one of the few who emerged to build that bridge. A young man of unusual talent, he managed to become the right person in the right place in multiple instances. As Kennedy wrote to him in 1985, "You had the right ideas at the right times and pursued them to perfection—always with elegant style. Hydraulics will always be in your debt for leading it to new planes of achievement and levels of elegance." With Rouse and others encouraging the invigorating, energizing merger of fluid mechanics and classical hydraulics, supplemented by healthy doses of the highly mathematical classical hydrodynamics and of small-scale model experimentation (which also developed first in Europe), hydraulics matured into a rigorous, predictive science based firmly on mathematics, theoretical analysis, experimental verification, and its original practical applications.

This critical merger occurred in large part through the mixing of bodies and minds across the Atlantic. In the late 1920s, U.S. graduate students started to travel to Europe to study in hydraulic-engineering laboratories, most drawn there through scholarships funded by the U.S. engineer John Freeman. Freeman, dedicated to the cross-oceanic exchange of hydraulic ideas, also arranged for European fluid mechanicians to lecture in the U.S., where eventually some of these scholars accepted positions. These events encouraged a rich cross-oceanic and cross-disciplinary hybridization of ideas among Europe's mechanical engineers, who were advancing fluid mechanics, and civil engineering hydraulicians in America.

The young Hunter Rouse's professional interests were gelling just when these European mentors were starting to influence American scholars. In 1929, fresh out of school with a bachelor's degree in civil engineering, Rouse had obtained an MIT fellowship to study for two years in Germany under Theodor Rehbock, who played a major role in the acceptance of the small-scale hydraulic model. Returning to the U.S., Rouse worked for two significant European fluid mechanicians, first at MIT as an assistant to Wilhelm Spannhake, then at Columbia University under Boris Bakhmeteff. In 1936, Rouse went to CalTech, where he had contact with Theodor von Karman, a renowned aeronautical engineer who had studied under Prandtl. By that time, Rouse had pinpointed his life's target: injecting the fluid-mechanical concepts that had proved so fruitful in aeronautics into the study of hydraulics. *Fluid Mechanics for Hydraulic Engineers* became both the title of his first book, published in 1938, and the driving passion of his life.

By the time he left CalTech for Iowa in 1939, Rouse had taught classes, written papers, and performed research that injected concepts of fluid mechanics into hydraulics, becoming one of the first to do so and thus placing himself at the forefront of such activities. He followed the same course at Iowa, initiating the first fluid-mechan-

ics classes and teaching laboratories here and continuing his research in diverse areas including turbulence, cavitation, sediment
transport, and boundary layers. Little could he have predicted how
the U.S. entry into World War II would intensify his fundamental
research. During the next five years, the Defense Department came
to IIHR, searching for information about the drag of Navy ships,
cavitation around torpedoes, fog dispersal in airports, and jets of
water to quench fires—fluids-related research topics that previously would have seemed out of place in a hydraulics laboratory.
However, Rouse and his IIHR colleagues welcomed funding for
these and similar projects and, in doing so, deepened the institute's
involvement in fundamental fluid mechanics. The payoffs were
great: generous funding, new equipment, and novel research initiatives. "A few [hydraulicians] became so deeply involved in experimental studies for the war effort that their professional lives were
never the same again," Rouse later wrote. This was true not only of
IIHR researchers; the entire institute was transformed during these
watershed years.

The war's aftermath also proved a bonus, for wartime research
had convinced the federal government that basic research would
more than pay its way in the solution of practical problems. Federal
research funding thus remained high and was channeled into fundamental studies such as fluid mechanics. The number of institute
researchers and the size of the budget soared. Rouse once again
serendipitously found himself in the right place at the right time.
Even during the quiet years before the war, several research initiatives that would continue to dominate IIHR had been established:
flow in pipes and drains and culverts, hydrology, sediment transport, fish-bypass investigations, and turbulence, the river studies
and small model studies that from the start had been an IIHR mainstay. Now, through newly formed programs such as those of the
Office of Naval Research and the National Science Foundation, the
institute's initiatives were joined by war-generated directives in
cavitation, diffusion of gases, pressure distribution on model buildings, jet characteristics, and, later, ship hydrodynamics. Essentially
being given carte blanche to follow his professional instincts, Rouse
rode without pause the surge of a federal funding wave that carried
him and the institute deeper into teaching and research on fundamental fluid mechanics–related facets of hydraulics.

This wave carried him through most of the following 20 years that he directed IIHR and freed him to expand his influence in multiple arenas. In this period, Rouse continued to instigate new classes in the field of fluid mechanics and to teach the old, integrating fluid mechanics firmly into the curriculum. He wrote textbooks—some that became classics—that allowed others to do the same elsewhere. He designed and exported teaching laboratories and equipment. He produced educational films. He hosted seminal professional conferences. He advanced the field by continuing his own research and writing, and he trained 35 doctoral and 45 master's-level graduate students to do the same. He took up the study of the history of hydraulics and became one of the few to write about the subject, again producing classic texts. He attracted quality students and colleagues who, as a body, expanded the institute's research base and magnified its size and the quantity of research support. As a widely sought consultant and lecturer, he traveled many times around the world in his efforts to spread his insights and foster creative relationships. He served on numerous governmental advisory boards. And in between, he administered first the institute and then, in later years, the College of Engineering, infusing both with his energy and ideas.

It did not take long for these efforts to spin a web that embraced the globe's hydraulics community. Rouse's well-known texts, films, and laboratory designs, exported abroad, became the tools to teach fluid mechanics and expand the curriculum of many a foreign school. They also exported Rouse's name, as did Rouse himself during his many trips abroad to lecture, consult, and keep in touch with colleagues. Foreign scientists were drawn to the institute to work or visit, gathering here ideas for expanding their own facilities back home. Students started to migrate to Iowa from around the world to study with the "master." Returning to their homelands, they took along Rouse's ideals and the mark of his demanding training and often became recognized for these in their own countries. Thus, Rouse was praised not only for his efforts in themselves but also for his fostering of international cooperation and information exchange in hydraulics. Through his efforts, contacts, and reputation, the stature of IIHR was expanded to the international scale.

Looking at these accomplishments, one is likely to think (as Kennedy quipped) of IIHR as "Rouse's lab." However, the reader would be mistaken to assume that all was done by Rouse. Rouse gave his life to hydraulics, identified with IIHR, and remained a strong leader with a controlling hand, but his laurels and the stature of the institute rest on others as well as him. Rouse was very much dependent on the talents of institute staff members, each of whom was teaching and accomplishing research achievements of his own. John McNown headed up post-war cavitation studies and took a major part in turbulence investigations. Emmett Laursen directed the many post-war sediment studies, becoming expert in scour around bridge piers and sediment transport. Lucien Brush studied sediment transport a few years later. Phil Hubbard's skills excelled in developing electronic instruments for fluid-flow measurement, Eduard Naudascher's in studies of hydroelastic vibrations. Chesley Posey was involved in general hydraulics for over three decades. Enzo Macagno came to the institute as a fluid mechanician but Rouse encouraged his interests in historical studies, which remain his active research field in 1997. Lou Landweber came to the institute in 1954 from a leadership position at the David Taylor Model Basin. Skilled in research as well as administration, he immediately took over the institute's large theoretical and experimental studies connected with naval architecture and hydraulics.

In academics, Professor Joe Howe was Rouse's closest confidant, sounding board, and greatest aid. Howe, who headed up hydrology at IIHR, was chair of the Department of Mechanics and Hydraulics throughout Rouse's directorship, and as such he took charge of instructional and academic matters. He shielded Rouse from the mundane administrative chores that Rouse disdained. Rouse, in charge of the research program, would discuss and clear important matters with Howe. The two developed a supportive, intimate partnership. "I never saw either acrimony or serious differences between the two," John McNown later wrote about his days at IIHR. "Each led one part of the activity and offered full support in the other." Howe also reviewed nearly all of Rouse's writings and co-authored *Basic Mechanics of Fluids*. Dale Harris, foreman of the institute's shop throughout Rouse's tenure, became expert at translating mechanical concepts into physical structures, thus producing unexcelled instrumentation and apparatus. He was the invaluable right-hand man when Rouse was designing teaching laboratories and research equipment, and also when Rouse was creating makeshift scenes for his films. These and the many other faculty and staff members were tied together socially as well as professionally. With relatively few staff and students, the Rouses decided to hold annual Christmas parties in their home. They regularly hosted dinner parties. For a period, monthly institute potlucks were followed by square dancing in the women's gymnasium, and students and staff ate and swam together regularly at summertime picnics. Sometimes the staff would pick up and take off for dinner together at the Amana Colonies.

Yet while Rouse could not have functioned without IIHR's staff, and while the staff as a whole advanced IIHR's goals, Rouse was the one who received the laurels. And these he received in abundance, as a listing of even his major awards exemplifies. For his influence on teaching hydraulics, the American Society of Engineering Education presented him the George Westinghouse and Vincent Bendix awards in 1948 and 1958 respectively. In 1966, he became one of the first 100 members of the National Academy of Engineering. A few years earlier, he had been granted the ASCE's Theodor von Karman medal. In 1979, the ASCE established a Hunter Rouse Hydraulic Engineering Lecture in his honor. Early in his career he had received prizes for his writings. In 1971 and 1991, IIHR further paid homage by publishing volumes of *Selected Writings of Hunter Rouse,* which reflect the diversity and elegance of his writings and also contain detailed sketches of his life and efforts. His historical efforts earned him the ASCE's History and Heritage Award in 1984, and the next year a symposium, "Megatrends in Hydraulics," was held in his honor at Colorado State University. Finally, when he was 85, the John Fritz Medal, an award for notable scientific or industrial achievement, was presented to Rouse "for pioneering the application of fluid mechanics to hydraulics, fusing theory and experimental techniques to firm the basis for modern engineering hydraulics."

One might wonder what personal attributes could earn someone these many accolades. Extremely talented, organized, focused, hardworking, rigorous, precise, thorough, demanding, critical, relentlessly tenacious, regimented, exacting, controlling, strict—all these traits describe Rouse. Insisting on excellence, he set extremely high goals for himself and for others. For example, any excess ink on the weblike nets of hand-drawn flowcharts sprinkled through his texts were whittled into perfection with a razor blade by Rouse himself. He knew what he wanted, and his days were devoted relent-

Hunter Rouse became known and respected at home and abroad for his research, teaching, educational innovation, and writing. Winner of many prestigious awards, he enhanced the stature of IIHR around the world.

lessly to achieving his goals. Rouse was a man with a mission. He lived with a single-minded focus on hydraulics and was driven to relate everything in his life to it. Frivolity and trivial efforts were unacceptable. Small talk was not pursued. Family vacations always incorporated a tour of some laboratory or other, a lecture given here or there. Unlike his predecessor Floyd Nagler and contemporaries such as Joe Howe, Rouse did not involve himself in community affairs. Associates in other departments found it difficult to talk to Rouse about anything but hydraulics. They also saw him as stern and serious, although some perceived a dry sense of humor underlying his words.

Personality can be read in part from appearance. At their 1929 graduation from MIT, Rouse and his classmates had been admonished to "dress, speak and act like a gentleman. . . . Keep one suit of clothes pressed every week. Never buy shoes unless you buy shoe trees for them. Keep them shined, shave yourself, and never wear the same collar at night which you wear all day." Rouse, a self-proclaimed perfectionist, took these dictates to an extreme. He always dressed formally for work, with a tie, his white shirt pressed to perfection, his collar starched exactly according to his stipulations. A suit jacket and long-sleeved shirt were the norm. His hair, the part perfectly straight, was cut weekly by the sole barber in town who met his standards. In clothing and appearance as in other matters, Rouse expected others to meet the standards that he imposed on himself. Even when students were working over the flume on a hot, humid, summer weekend, Rouse sent them to "go and get decent" if they dared to strip off a shirt or to appear in shorts. Students learned quickly what was expected. They dressed well for the Christmas party at Rouse's house, often purchasing new clothes for the occasion.

Rouse expected the running of the laboratory to follow similar high standards. In this way, IIHR was indeed "Rouse's lab." He was the one in control. He monitored the smallest details, tracking students and employees, expecting them to accept his goals as their own. The lab was to be kept free of clutter and mess. Students were to dedicate themselves to their work, and second-rate efforts or products were unacceptable. The lab was run in hierarchical fashion, with Rouse making decisions and passing them on to the next in command. Rouse only spoke directly to the shop foreman, for example, even if those who were to be working on a project were also present. Rouse assumed that things would be done his way, and students learned a tremendous amount by following this dictum. However, some found his Teutonic approach too strict, too controlling and confining. Thus, while Rouse's demands raised students to levels they didn't know they could reach, these demands tended to stifle creativity. With Rouse so clearly and strongly at the helm, his fellow academics were not allowed the latitude to follow their own leads and rise to their own full potential.

Although Rouse's high standards challenged others to do their best, they also created problems. Rouse appeared cold to some. His colleagues recognized his unusual gifts, talents, and contributions, and for the most part overlooked his human foibles. Yet Rouse's extremes separated him from others. Some students, although admiring Rouse tremendously and acknowledging him as the pinnacle of their professional education, clearly feared him. In place of love, he received respect; instead of warmth, he drew awe. His many accomplishments undoubtedly were achieved with tremendous satisfaction, but also at a price.

Rouse's accomplishments are described further in the following chapters, but two traits deserve additional note here. The first was a strong love of writing and the written word. Perhaps this started with making up stories for the puppet shows he presented as a child. By his college years, it was firmly entrenched. As an engineering undergraduate at MIT, he won prizes in English composition, tutored foreign students in composition, wrote skits, and worked on the student newspaper, eventually becoming its editor. By his late twenties, he was working on his first text, which was to be followed by several other books. These mirror in content the flow of his life: strict fluid mechanics and research articles in early life yielding to articles on education and administration during his later years at Iowa, both gradually mingling with historical articles and books and a few philosophical treatises as he aged. He never quit writing. After retiring, he published articles on his hobby of lapidary. He obviously relished expressing the wanderings of his mind through words.

Rouse's prolific writings as a whole reflect the way he lived: grammatically exact, executed to perfection, precise, complex in sentence structure, formal, flowing and seamless, polished, eloquent; logical and consistent, to the point, extremely detailed, seeking new information with delight, but lacking in metaphor, paradox, or ambiguity.

With writing, as with other matters, he expected his students to meet his high standards. "Above all, [the engineer] must be able to communicate: the English language should become as useful a tool to him as mathematics," he wrote. When critiquing theses, Rouse would place an "X" in the column where he noted any error of content, grammar, or form, but would never explain the nature of that error. That was for the student to figure out. Students would gather in groups to puzzle for hours over these enigmatic markings.

Several of his articles report on his foreign travels and his thoughts of foreign cultures. These reflect another major intrigue, his love of travel and broad fascination with foreign lands. The amount that Rouse traveled was indeed astounding. His wife calculated that he spent half of his life away from home during the 1940s and 1950s. His mother tallied the miles and noted that by 1967, he had traveled half a million miles by air. One or two dozen trips a year were not unusual, with destinations ranging from Chicago and Wisconsin Rapids to Istanbul, Bora Bora, Singapore, Leningrad, or Perth.

None of these trips was frivolous. Although his family sometimes would join him, all his travel was work-related. He went elsewhere to present papers, to consult, to examine research laboratories, to meet with colleagues. He met with former graduate students who would invite him to visit and lecture in their homelands, and he actively recruited new international students. He spent 1952–53 at Grenoble as a Fulbright Scholar. A year in 1958–59 also was spent in Europe, this time as an NSF fellow, during which time he earned a second doctorate at the University of Paris. He suggested that the U.S. State Department arrange an exchange of American and Soviet hydraulics directors, and in the early 1960s he was one of five Americans to participate in this exchange. In 1965, he headed an American delegation to Japan for a seminar on instrumentation in hydraulic research. In 1974, he toured the People's Republic of China as a member of the first group of American engineers to enter that country under its current political regime. This sampling of Rouse's foreign experiences exemplifies why he was eulogized for tirelessly fostering exchanges of personnel, information, and international goodwill. Rouse clearly valued experiences in other lands and other cultures, both for himself and for others. He boasted that about half of the institute's graduate students had come from abroad, and he defended international education as beneficial to all parties involved and a boost to the world's stability as well as America's strength. He encouraged foreign experiences among IIHR staff members as "invaluable for the professional, social, and political understanding which [they] stimulate."

Rouse, through the years, had steadfastly avoided sacrificing his love of research and teaching for administration of any sort, which he considered to be a lower endeavor. He did not seek or hold elected offices in professional societies. He insisted that no one in his right mind would consider an administrative position such as that of dean, and he encouraged others to reject such positions in favor of continued research. Thus, it may have surprised him as well as others when he found himself accepting the deanship of Iowa's College of Engineering late in 1965. Circumstances had forced this position upon him. Dean Melloh, who had come to Iowa in 1960, bred such discord and tension among the engineering faculty through his administrative policies that he was dismissed by the university's President Bowen in 1965. The college was left in such turmoil that "no desirable candidate . . . could be attracted from outside [the university], and I hence accepted President Bowen's appointment to the end of improving the situation as much as possible in the time remaining at my disposal," Rouse later wrote. This was not Rouse's only problem. In addition, the number of students enrolled in engineering was dropping, and federal support of science in general was dwindling.

Colleagues later commented that as dean, Rouse seemed like a fish out of water. Yet once installed, he flung himself with characteristic dedication into rebuilding the college. During his six years as dean, and with the help of able faculty and staff, Rouse guided a complete revision of the undergraduate curriculum. He directed the remodeling of much of the Engineering Building and renovated the college library. He reinstated communications with alumni through mailings and symposia and instituted a Dean's Gift Fund to pay for these efforts and scholarships. Rouse convened an ongoing College Advisory Board as well as a college faculty council. Rouse's strengthening of the college faculty as a whole produced a quadrupling of the annual publication rate (to about 250) and tripling of the research support (to $1.2 million), removing IIHR from the clear domination it had held six years earlier (when IIHR had produced nearly a third of the college's total publications and brought in about 40% of its research support). Meanwhile, he continued his writing, shifting the focus of several articles to expositions on education: in Germany, in the People's Republic of China, among foreign students, in fluid mechanics, at Iowa. Neither his domestic nor his foreign travels slackened.

Engineering-student enrollment continued to decline, following a national trend and "a nationwide reaction against technology." Rouse mounted a hearty recruitment campaign in response, consisting of radio broadcasts, visits to high schools, and printed materials—efforts that were largely unsuccessful.

During these years, his declarations of his exacting standards and beliefs remained firm and loud. A main and oft-repeated theme was the university's function as both producer and disseminator of knowledge. This meant that teaching and research were each to occupy half of a professor's energy, for "the two are together necessary to a proper academic balance. Neither is sufficient in itself." Rouse wrote that if one was incapable of doing both well, "he does not belong in a university." Rouse also assumed that student behavior should and would remain within societal norms—and in general it did at Iowa. In 1969—when tear-gas canisters were exploding among crowds of incensed student demonstrators across the country—Rouse reported that "engineering students have shown no tendency to move in the minority direction—aside from a few beards and sideburns."

Yet while Rouse's ideas remained firm, he attempted to step back and allow others to take the lead. The need to rebuild consensus among college faculty was an effort for a person as authoritarian as Rouse. Nevertheless, he later wrote that "seeing a general desire on the part of the faculty that I handle all matters democratically, I made every effort to do so, despite my belief in the greater efficiency of the autocratic process (and the fear on the part of some that I would be a dictator)." In his later writings, one even can detect a glimmering of the mellowing that comes to many with age, a slacking off from the desire for absolute control and correctness. In 1976, the year he retired from the University of Iowa, Rouse wrote the following in a critique of the management of research institutions:

> There was a time, many years ago, when I thought that I knew just what course the world as a whole should take, and I suppose there was a period thereafter when I had some slight influence upon it. But planning effectively for the future is a responsibility of the young and early middle-aged, not the old. One acquires inhibitions after one matures, and progress needs unfettered strides in new directions. . . . There are, on the contrary, many different ways of doing almost anything—some not so good, to be sure, but enough of others to satisfy particular situations and the peculiarities of the individuals involved.

Rouse had accepted the deanship to "lead the College to new levels of excellence." In 1972, feeling that he had accomplished this goal, he thought it time to step down. He retired to an office on the fifth floor of the Hydraulics Laboratory, where he wrote yet another book, *Hydraulics in the United States 1776–1976*, which was published in 1976, the nation's bicentennial. Retirement from the University of Iowa took him for a decade to summer teaching positions at Colorado State University and winters spent in Arizona, where he worked with gemstones and crafted jewelry, also shaping his thoughts into articles on lapidary. He continued to travel, write, present guest lectures, and write articles, now most often on historical topics, until Parkinson's disease limited his activities and tied him to Sun City, Arizona, where he died at the age of 90 in October 1996.

7. Research, 1933–66

In the late 1930s, researchers at IIHR delved into the problems of plumbing fixtures that drew wastewater, water-hammer noises in pipes, and pollution-spreading cross-connections of indoor water systems. While these topics may seem mundane today, improperly designed plumbing systems were a major concern then. And improperly designed they were, commonly enough that IIHR received considerable research and testing funding from a consortium of plumbing manufacturers, the National Association of Master Plumbers, and was named the National Plumbing Laboratory's official testing center, a position that afforded IIHR considerable stature and repute. These very applied plumbing efforts coexisted alongside hydraulic model studies funded by the U.S. Army Corps of Engineers (COE) and a few unfunded basic-research projects.

All this was to change in the following decade, when the Office of Naval Research (ONR, founded in 1946) replaced the COE as the institute's chief governmental supporter. At that time, applied efforts retreated in relative importance and fundamental studies stepped into the limelight. Wind and water tunnels were added to IIHR's stock of standard equipment. Broad-scoped grants from the National Science Foundation (created in 1950) were sought in place of very specific contracts from the Master Plumber's Association or similar organizations. The tenor of the institute shifted dramatically, reflecting an evolution of society's needs and expectations—but also shoved forward by World War II and the firm convictions of Hunter Rouse, the institute's new and dynamic director.

The shifts displayed so dramatically at IIHR were one small scene in the nation's picture of altered research priorities, which is perhaps best envisioned by examining research funding in this country. While ideally the minds of scientists and engineers cut their own creative channel, in reality the choice of their investigations follows a course determined by finances, which in turn reflects the societal needs and constructs of the time. Early in the 1900s, these needs and constructs brought industry and university research together in productive symbiosis: industry providing the funding, universities providing the facilities, the intellect, and a continuous supply of well-trained personnel. This symbiosis was responsible for the creation of many large university-based laboratories, most of which are a product of the twentieth century. Government moneys for research remained extremely limited before 1940, except for a brief surge of defense-related funding during World War I, which plummeted as soon as the war was over. The meager government sponsorship that remained was pumped largely into agriculture.

Industry, in contrast, started in the early twentieth century to embrace wholeheartedly both applied and fundamental scientific and engineering research for one reason: there was money in it. Recognizing that their continued profits depended on new ideas, larger industries created their own in-house laboratories. But in-house labs were expensive to maintain and thus were often out of reach for the smaller industries. Research institutes associated with universities offered a solution. Industries that sponsored such institutes enabled universities to keep up with the needs of manufacturers and provided manufacturers with expertise when needed. Thus, early-twentieth-century industry, for the first time, actively supported and cooperated with university science and engineering departments, and university laboratories became essentially part of an all-embracing industrial system. Increasingly, engineering educators and industry leaders cooperated in establishing joint research programs throughout the United States.

The Hydraulics Laboratory was constructed in the 1920s with this industry-university symbiosis in mind. Dependent from the start on outside funding, Floyd Nagler initially equipped the laboratory as a testing center for the bulky turbines of hydroelectric plants. He felt that such a program would guarantee its own success, since no turbine-testing facilities existed outside of Massachusetts. While this endeavor never came to fruition, Nagler did

commence other industry-funded projects such as an intensive program for testing current meters, which measure river flow.

Starting in the late 1930s, the plumbing research mentioned earlier became the centerpiece of IIHR's industry-oriented research. At
that time, faulty plumbing fixtures were commonly manufactured
alongside those that functioned well. The result could be poorly designed water-supply systems that pounded out water-hammer
noises, or that back-siphoned toilet water or (in hospitals) formalde-
hyde-contaminated water from specimen tanks into water-supply
pipes. The institute thus became the National Plumbing Laboratory's unbiased scientific agency to test plumbing equipment and
investigate the hydraulics of plumbing systems. These contracts
were Dean Dawson's gift to the institute when he arrived from Wisconsin as a hydraulics professor and an established expert in
plumbing, with ongoing plumbing projects and numerous industrial connections. In Iowa he, along with Anton Kalinske (who had
followed Dawson to Iowa from Wisconsin), continued to perform
multiple studies to determine which systems were safe from back-
siphonage and how to design piping systems, pipe-venting requirements, and other features central to the safe and efficient delivery of
water in buildings. They investigated the flow of air and water in
vertical pipes, the hydraulics of flush toilet bowls, the intrusion of
sewer gas into buildings, and the structure of effective grease traps.
Their plumbing efforts addressed fundamental questions as well,
such as the mixing of air and water in vertical flows and the formation of vacuums in pipes.

For all these studies, and also for research and instruction in
laminar and turbulent flow, pipe resistance, and other fundamental
studies, circular pipe systems were constantly being constructed,
torn down, and refigured at the institute. Sometimes, to allow better
visualization and photography, these pipe systems were fabricated
from transparent plastic, an uncommon practice at the time and a
source of pride for the institute. Prolific plumbing-related publications by Kalinske and Dawson resulted, a tribute to Dawson's dedication to hydraulics despite his administrative duties. These studies
demonstrated the importance of good plumbing for public health.
The results of studies were expressed in numerous ways, some not
traditional for hydraulicians—for example, through papers in the
American Journal of Public Health, two educational films on plumbing

Investigations of the hydraulics of flush toilets and other plumbing matters extended the efforts of IIHR researchers into everyday life in a visible, practical manner. IIHR stipulations determined permissible drainstack and vent sizes, standardized design criteria for water delivery systems, and set plumbing codes and governmental regulations around the country.

and public health that were made at IIHR and shown in practically every state, and (in the late 1940s) demonstration lectures to the university's medical students on plumbing dangers and sanitation.

Other industry-funded projects in the 1930s and 1940s included the institute's product-development and commercial testing program, which focused on specialized devices and tests of a non-routine nature. These tests were viewed as important to staff and students because the tests demanded the application of formal knowledge to very practical problems. The plumbing project included the testing of hundreds of valves, hydrants, vacuum breakers, and other plumbing devices for manufacturers. Testing standards were also developed for kitchen and restaurant grease interceptors, an effort expanded during the war to the testing and rating of grease interceptors for large army kitchens. The institute then helped many commercial firms revise their designs accordingly. Similar efforts included the testing of plastic tubing and fittings for Dow Chemical, joint closures for airplane fuel lines, automatic valves, and flow meters.

While Floyd Nagler had assumed at first that industry would be his laboratory's primary funder, that belief was soon dispelled by Nagler's fortuitous partnership with David Yarnell, an engineer employed by the U.S. Department of Agriculture (USDA). Yarnell was sent to the Hydraulics Laboratory in 1921, soon after its opening, to perform a series of full-scale tests of culverts. Yarnell and Nagler quickly developed a friendship and a productive partnership that persuaded Yarnell to remain at the Hydraulics Laboratory. He continued to bring an active program of USDA-sponsored research projects to its doors until his death in 1937. Yarnell supported a number of graduate students as engineering assistants for these projects. He co-authored many papers with Nagler, became an associate director of the laboratory, and with Nagler actively advocated the lab's expansion. Through him, IIHR became the recipient of U.S. government funding decades before such support of university research was a common assumption.

This became the first of a series of cooperative agreements through which government agencies brought their staff to the institute to perform research. Such efforts ranged from temporary projects with single federal employees to large federal sub-offices with up to 40 staff members who carried out a continuous program of experimentation. The projects utilized the institute's space, equipment, and expertise, in return feeding the institute with a dependable source of funds and equipment. Their rental fees were largely responsible for repaying the construction costs of Nagler's expansions of the Hydraulics Laboratory. Institute staff benefited from contact with the federal employees and at times contributed facilities and staff time to these projects. In the 1930s, research performed through these agreements accounted for the vast majority of the river hydraulics, hydrology, sediment measurement, stream flow, and model studies performed at the institute.

The dominant cooperative program was that of the COE, who had opened a sub-office in the lab in 1929 under the direction of Martin Nelson, another government engineer who (like Yarnell) soon became an associate director of the institute. The COE was starting to construct a 9-foot channel in the upper Mississippi to improve navigation and reshape the river into a major transportation corridor (see chapter 3 for Floyd Nagler's involvement in this effort). It decided to perform an extensive series of model tests for the dams, locks, and spillways that would be constructed as part of

Cooperative arrangements with the COE brought dozens of studies and staff members to the Hydraulics Lab to model dams, spillways, and locks for a 9-foot upper Mississippi River channel.

that project. The model program quickly filled the newly expanded laboratory and continued to grow. In the 1930s, the program dominated the institute in size and in number of people as well as finances. Between one and two dozen studies might be carried out in a single year, occupying as many as 30 COE staff who were stationed at the institute—the majority of researchers present. Dozens of lengthy technical reports resulted, each contributing a portion to reshaping the Mississippi into what we see today.

In all, models of locks for 15 locations above St. Louis were tested here. Navigation structures for the Ohio, Illinois, Tennessee, and several other rivers were also modeled. Sometimes detailed structures of specific locks and dams graced the laboratory; at other times, miniature versions of long river reaches flowed through the second floor of the institute. These studies demonstrated that by the 1930s, small-scale physical modeling was accepted as the crux of hydraulic research. Its prominence at the time was hailed by Paul

Thompson, director of the Waterways Experiment Station in Mississippi, in a talk he presented at the institute in 1939: "[Hydraulic model testing's] importance now is taken for granted. The small-scale model has become an accepted and proven aid to the designing engineer, and as such it is entitled to the prestige, dignity—and anonymity—of, say, the slide rule." The ability to visualize fluid phenomena in models was recognized as important to their interpretation; thus, starting in 1935, glass and transparent plastic flumes started to appear in the lab, and these were filled with clear water rather than the murky liquid pumped up from the Iowa River.

COE funds, provided largely through the federal government's WPA public-works program to reduce unemployment during the Great Depression, kept the institute's workers alive during the financially stressed 1930s. In fact, it could be said that this cooperative agreement kept IIHR afloat during the difficult decade following Nagler's death, when a seeming dearth of spirited leadership weakened IIHR internally at the same time that the nation's economy assaulted such research institutes from the outside. Mavis's trepidation at the loss of such funding echoes through a letter he wrote on February 28, 1936, when he realized that the facilities of the St. Anthony Falls Hydraulics Laboratory in St. Paul, then under construction, could overshadow the institute's laboratory and lure away IIHR's federal moneys. "During the past few years hydraulic engineering research at . . . Iowa has been possible only because [COE] income was available to support the research and to make essential purchases of supplies and laboratory equipment," he wrote to the dean. "If the support of the federal government were withdrawn . . . , my half-time secretary and I could arrange between us to answer the telephone." Luckily the COE maintained its IIHR sub-office until space constraints forced its removal in 1948, and Mavis's fears were never realized.

Cooperative agreements with other federal agencies were smaller in size, brought fewer staff to the institute, and provided less funding, but were very evident around the institute during the 1930s. USDA staff at first engaged in studies of culverts, highways, and bridges and then studied the flow of water in conduit bends, sediment transport, and hydrology. When Yarnell died in 1937, the USDA's presence was assumed by the Soil Conservation Service, which supported the continuous monitoring of the Ralston Creek watershed until 1988. The U.S. Weather Bureau stationed a branch office of the flood-prediction service at the institute between 1938 and 1942. This organization focused on developing more scientific methods of flood prediction, thus foreshadowing a significant mission of IIHR's 1990s research program. Its tests were responsible in 1938 for pulling the Rapid Creek watershed into the long-term monitoring studies that were already being performed on Ralston Creek.

The U.S. Geological Survey (USGS—Water Resources Branch, Surface Water Division) established a district office at the institute under Mr. Kasel in 1932, which remained at the institute (albeit under different leadership) until 1968. Its heads, like those of the USDA and USGS cooperative agreements, became associate directors of the institute. The USGS office measured the stage and flow of streams throughout Iowa and also participated in hydrologic and sediment studies. And last, the COE, USDA, and USGS, jointly with three other federal agencies (Bureau of Indian Affairs, Bureau of Reclamation, and Tennessee Valley Authority) and the institute, started in 1940 a multiple-year, extensive study of sediment measurement and analysis. This study was performed under Associate Director Lane's direction and was the culmination of his research career at the institute.

While most pre–World War II funding came through such cooperative arrangements, a few government-sponsored contracts were awarded to the institute. The Iowa Conservation Commission, for example, funded several years of fishway studies in the late 1930s that became a harbinger of a massive salmon research program (see chapter 15). Floyd Nagler, shortly before his death, had complained about the waste and inefficiency of fish-bypass systems in Iowa's streams. By 1937, institute personnel were reviewing the world's literature on fishways, performing research on model fish ladders of various designs, and testing full-sized fishways in the large basement channel. In effect, these fishway studies also became investigations of migration of Iowa's fish. In the final project report, facts about fishways were mingled with observations about the upstream movements of quillback, carp, and catfish, such as the dates of migration (mostly between May 15 and July 15) and temperatures (not much happened when temperatures dropped below 65°F).

This sediment measuring device, about to be lowered into the river, was one of many sediment samplers that IIHR tested for their ability to assess sediment size and concentration at all stream depths. A decade later, the project to assess various sediment samplers was described as one of IIHR's most comprehensive studies of all time.

With most of the pre–World War II funding coming from industries and government agencies with specific problem-oriented agendas in mind, it is understandable why IIHR's research was dominated by applied rather than fundamental studies. This was the mandate and the expectation of the time. As Nagler himself wrote, "The chief purpose of the research hydraulic laboratory is to solve problems in which the prosperity of Iowa is involved." For most of Nagler's tenure, the engineering college had remained the College of *Applied* Sciences. During the following decade, in the 1930s, institute research was categorized according to its practical applications: flood control, pier obstructions, weather studies, dams, measuring devices, flow in closed conduits, head regainers,

IIHR's attempts to develop effective fish passageways commenced in the late 1930s, when 15 species of fish—and a total of 13,000 fish in one year—were induced to climb various experimental fish ladders.

and the like. The annual reports from this period stated the case quite clearly: "The IIHR was organized to carry on . . . [cooperative research for] agencies which have *hydraulic problems requiring solution*" [emphasis mine]. This narrow, project-oriented categorization of research efforts was dictated by that period's need for straightforward answers to very practical questions.

Thus, it is to the institute's credit that any fundamental studies at all were performed in the nooks and crannies that remained after government agencies and plumbing fixtures had claimed their territories. Yet some fundamental research was executed in the 1930s, primarily in the fields of sediment transport, hydraulic jumps, and turbulence. The formation and characteristics of hydraulic jump—the process whereby a rapid, shallow flow suddenly becomes tranquil and deep, and kinetic energy is transformed into potential energy or dissipated—were studied by Yarnell, Lane, and Posey, in both open channels and within pipes. Tests were also conducted on the resistance of channels and pipes, as was an extensive eight-year project on the conversion of kinetic to potential energy in various pipe expansions.

Turbulence investigations had been initiated in the late 1920s by Yarnell and Nagler. They studied the effect of turbulence on the registration of current meters and won the J. James R. Croes Medal for their paper on the subject in 1932. Turbulence research was resurrected in 1936 by Kalinske, who played a major and energetic role in both applied and fundamental investigations at the institute and expanded turbulence-research efforts until he left the institute after the war. He worked with a number of graduate students on turbulence-related thesis topics. Rouse likewise delved into turbulence studies as well as fluid mechanics more generally. During the war, the understanding of turbulence took on heightened significance for many defense-related projects, for example measurements of fluid drag on various bodies and the separation of grease from water.

Studies of sediment's relationship to turbulence and of the percolation of water through sediment were commenced by Lane and Kalinske. Their long-term program on the laws governing transport of suspended sediment provided topics for a number of theses and was picked up by Rouse after he came to IIHR. These fundamental sediment studies interplayed with the institute's ongoing applied sediment studies that were funded by government agencies and that focused more on measurement and analysis of coarse bed load.

Some of this fundamental research work was funded to the tune of a few hundred dollars by the Hydraulics Research Committee of the American Society of Civil Engineers (ASCE) or the Engineering Foundation, or (when performed by graduate students) by the Graduate College or the Mechanics and Hydraulics department, but in general fundamental work was financed by the institute without funding from any outside source. Moneys for facilities and staff were then, and have remained, largely generated by the institute (rather than financed directly by state appropriations), and through the 1930s the vast majority of funds were generated from the river-model studies of the COE. During the 1939–40 fiscal year, for example, the COE injected $8,895 directly into the institute's budget—providing 72% of the $12,258 total for the year. A few years earlier, in 1935–36, the COE had provided 95% of the institute's income. These were the funds that gave the institute its financial wherewithal and allowed its few faculty members and their graduate students to complete the above-described fundamental research projects. In addition, the COE annually spent about $50,000 on salaries, labor, and materials for their IIHR modeling efforts. These projects provided the necessary equipment and also the trained personnel that gave the institute a critical mass and maintained creative discourse.

Thus, in 1940, when the institute was threatened with the loss of these funds, it feared the virtual closing of its doors. "Plans are under way to discontinue [the COE's] work here," stated the institute's Annual Report for 1940. "This will eliminate the greater part of the income of the Institute and when the balance on hand is exhausted, will necessitate a practically complete cessation or very great curtailment of nearly all of the activities of the Institute. . . . It is therefore vital that some new source of income . . . becomes available."

Once again, national crisis came to the rescue. World War II demanded a response from all sectors, and in 1942 hydraulic engineers were called to serve their nation by applying their knowledge in new manners. For such applications, the government was willing to pay. While earlier the value of research might have been coldly assessed in terms of profit and efficiency, now—with the lives of soldiers and nation-states on the line—such questions became secondary. While a few years earlier the federal government's withdrawal was threatening to deliver the institute a death blow, now federal funds bestowed new life. Numerous grants, coming largely through the National Defense Research Committee (NDRC), enabled newly hired staff members working with newly constructed equipment to address diverse new questions and fulfill new directives.

In short, international tragedy restored the institute, administered a massive dose of growth hormone, and propelled its research

onto the fast track. Suddenly time replaced funding as the chief lim-
iting resource. Students dropped their classes and were hired as researchers. The pace of work heightened to a frenzy as workers remained late into the night to complete urgent projects. Methods
were needed to disperse low-hanging fog that obscured British
landing fields and threatened the safety of pilots returning from
bombing missions. Fire-fighting equipment had to be developed for
Coast Guard vessels whose monitors were unable to shoot far
enough a concentrated stream of water strong enough to quench
distant flames. Tiny bubbles forming around the speeding heads of
torpedoes had to be eliminated in order to increase torpedo
efficiency. These problems and others laid on IIHR's steps all had
their hydraulic components. Mississippi River models and plumbing fixtures were shoved aside. In their place, the institute built its
first pressurized water tunnels to handle experiments with cavitation and its first wind tunnels to deal with fog-dispersion and gas-
diffusion studies. These and other wartime research initiatives are
described in chapter 9.

The war's remolding of the institute was paralleled by transformation from within. Emory Lane, the institute's associate director,
felt he could best serve his country by taking leave of IIHR to assist
the Tennessee Valley Authority generate more power for the war
industry. Institute researchers Hunter Rouse and Anton Kalinske,
who were asked to step temporarily into Lane's position, were not
content to steer the same course as their predecessor. Their love was
the fundamental processes of fluid flow. The war offered ample opportunities to study basic hydraulic processes while engineering
the flow of water around a torpedo head or modeling air flow over
mountainous terrain. "Practically all problems posed by the various
war agencies were of a fundamental character," Rouse wrote in
1945. Rouse and Kalinske sought the wartime projects and obtained
the wartime funding described above, and then used these to boost
the emphasis on fundamental research at the institute. Two years
later, Lane returned to a transformed laboratory, one that had
grown far beyond his vision and capabilities. Rouse was invited to
assume permanent directorship of the institute. He firmly grasped
this opportunity and continued his efforts to remold the institute's
research program along the lines of fundamental principles, avoiding the routine applied studies, commercial development efforts,

and product-testing projects of earlier years. The war had thus handed Hunter Rouse the position, funding, and research initiatives to fulfill his professional dream: to build a research program solidly based on fluid mechanics rather than empiricism.

All this defense-stimulated innovation might have been for naught, had the government's research funding plummeted after World War II as it did after World War I. But with the commencement of the Cold War, quite the opposite happened. The wartime contracts set the stage for even larger and more diverse post-war governmental efforts at this university and across the country. Recognizing that research was becoming ever more expensive, Congress passed unprecedented appropriations for supporting research, and NDRC contracts were transferred to other governmental defense agencies without a wink. With these transfers, the federal government assumed a new role: it became the leading supporter of university research laboratories. Peacetime federal funds for the first time joined industrial sponsorship of these labs, and in some cases largely replaced private moneys. The federal government had finally accepted what industry had realized for many years: that funding of university research facilities paid off.

This new government funding went beyond the applied, production-oriented research that industry had sponsored. Because fundamental research had rapidly produced an abundance of strategic military applications such as radar and atomic energy, many of which could not have been predicted, the funding of fundamental research was now recognized as a legitimate governmental function in the national interest and for the public good. Wartime research had also displayed the bonds between fluid mechanics and applied hydraulics, with contract projects often providing valuable basic information on fluid behavior. As Rouse later stated, "The [Second World] War, in a word, had greatly accelerated the transformation of the hydraulics profession into a full-fledged user of and contributor to the deepening pool of fluid-mechanics knowledge." The government's new insights lined up exactly with the desires of the institute's new director: both pushed simultaneously toward expansion of fundamental research—a happenstance of amazing serendipity.

Following the war, most of the institute's wartime contracts were assumed by the Navy's Bureau of Ships. Contract wording was left purposefully broad so that "a wide variety of basic problems in hydrodynamics may be investigated as the need arises"—a factor that undoubtedly pleased Rouse. In addition, the Navy made clear from the start that it would provide long-term funding for the institute's fundamental research, and it supplied IIHR's mechanical shop with surplus equipment from the war, some of which is still in use. It also embraced and supported students working on contract projects, since the training of future research personnel matched the advancement of fundamental knowledge as the Navy's primary goal. A few years later, in 1948, the Bureau of Ships contract was expanded and transferred to ONR, which was the government's first (and at that time only) agency supporting basic scientific research at academic institutions.

ONR, founded in 1946, encompassed a remarkable diversity of research projects at universities across the country—from those concerned with superconductivity, dental caries, and turbojets, to investigations into computer technology, radio astronomy, and atomic physics. Some of these foci were later adopted by other new government agencies such as the National Institutes of Health, National Science Foundation, and Atomic Energy Commission, that were created after ONR and likewise concentrated on basic research. However, IIHR had developed a hearty relationship with the Navy during the war. While the institute's federal funding broadened with the years, its postwar bond to ONR and the Navy remained strong. ONR remained IIHR's single largest funder throughout Rouse's term, and for much of that time its initiatives strongly dominated the institute. At times, ONR funded nearly a dozen different fundamental projects simultaneously. Ongoing funding allowed researchers to assume an enviable continuity of long-term projects and freedom from financial insecurity. Until 1963, ONR consistently provided over half of the institute's budget. During those years, other Navy agencies combined chipped in another 10% or so of the institute's total budget, occasionally raising the total Navy contribution to over 70%. Although the relative domination of institute activities by ONR declined after 1963 and never again approached that of the immediate post-war period, ONR has remained a major institute research funder.

These Navy contracts sustained research in flow analysis. Chief among these were investigations into turbulence, cavitation, and

In the years following World War II, generous funding from the Office of Naval Research dominated IIHR's initiatives and activities. A portion of the assistance came as new equipment such as this wind tunnel, built in 1948 for ship resistance studies.

pressure distribution, efforts that continued throughout Rouse's tenure. Surveys of turbulence and its diffusion expanded into analysis of the development of boundary layers on various surfaces, jet characteristics, flow over rough and smooth surfaces and at abrupt expansions, turbulence of the hydraulic jump, and other manifestations of rough waters. Cavitation studies were expanded to experiments on submerged nose forms at various angles, and studies of submerged jets, their noise, and their surrounding eddies. Also investigated was the relationship of pressure distribution and cavitation around various forms. Cavitation studies, which had been commenced by Anton Kalinske during the war, were directed by John McNown, who with Rouse also took a major part in the turbulence studies.

These ONR studies formed much of the basis of the institute's fundamental research efforts during the 1940s and 1950s. However, they were not the institute's sole research efforts. The institute continued its wartime studies of fire nozzles and wind flow over mountains, as well as the hydrological and plumbing investigations that had pre-dated the war. The COE resumed and for a time expanded river-model studies. Other occasional projects included various studies of flow characteristics in open channels, the effects of screens and perforated plates on flow in closed conduits, characterization of pipe manifolds, wind pressure on structures, stratified flow and thermal mixing, and various river-control efforts including model studies of flow-control structures, dams and spillways, recirculating cooling-water systems, and the like. Some of these were carried out for the Engineering Foundation or ASCE or were contract projects funded by a city, engineering firm, or industry, sometimes here and sometimes abroad. Others were underwritten by the institute.

The resumption of earlier studies and continued execution of wartime efforts soon brought laboratory space to the crisis mode—so much so that the Hydraulics Lab's elevator shaft was converted to office space. With the luxury of abundant Navy funds, Rouse could afford to be picky. The plumbing studies were moved across the river into the Materials Testing Laboratory. Then Rouse made clear that the COE's "routine testing work," which had kept the institute alive in the previous decade, was no longer welcome. By the end of 1947, Rouse was able to report to the dean that under "mutual agreement," the COE had decided to remove their voluminous lock and dam models to a lab "more conveniently located." Institute facilities could thus "be devoted in full to the fundamental projects which it is best suited to conduct." The following year, the Model (later renamed West) Annex was opened across Riverside Drive from the Hydraulics Laboratory. The annex housed several pieces of equipment for sediment studies—a new recirculating transport flume, scour flume, and supplementary equipment to prepare bed material and analyze samples, all funded by ONR—as well as a wind tunnel and instructional laboratory. The combination of this new building and the space vacated by the COE allowed abundant room in the Hydraulics Laboratory for offices, an electronics shop, a lecture hall, an expanded student laboratory, and a new ONR-funded variable-pressure water tunnel. The elevator shaft was retrofitted with a new elevator, and it once again assumed its original purpose.

The Model Annex was devoted nearly exclusively to sediment studies, which were funded by ONR's broad new 1948 contract. For nearly a decade, this funding greatly expanded the institute's previously unfunded queries into theoretical aspects of sediment behavior—investigations into the mechanisms of entrainment and transport, settling rates of variously shaped particles, measurement of their size-frequency distribution, and rates of scour. A number of graduate theses resulted from these efforts. During the same period, major investigations of scour around bridge piers and sediment deposition in conduits and traps were funded by the U.S. Bureau of Public Roads and the Iowa State Highway Commission, the only other significant funders of institute research during the late 1940s and early 1950s. Sediment studies captured increasingly more interest and (with turbulence and cavitation) dominated research here until 1956, when ONR sponsorship of sediment work was terminated because such research "could be sponsored more appropriately by another organization." A few years later, in 1959, the bridge scour and conduit studies culminated in the publication of manuals with design criteria for bridge piers, abutments, and storm drains for sediment-transporting flow, as well as suggestions for alleviating scour. A year earlier, Emmett Laursen, who had directed these studies, had accepted a faculty position at Michigan State. These changes effectively terminated IIHR's intense focus on sediment for the remainder of Rouse's tenure, although the subject still fostered occasional graduate theses, contract efforts, and National Science Foundation grants.

The institute's output in the post-war period was not limited to research. Laboratories were designed and built to be installed in other countries, educational movies were made for students and consumers, hydraulics conferences were organized to bring professionals together. Texts were written and historical studies performed, efforts described here in chapters 10 and 11. In addition, starting in the late 1940s, the institute began developing electronic instrumentation, another research effort funded by ONR. The institute's cutting-edge research demanded cutting-edge techniques for measuring fluid flow and pressure. Prior to World War II, most measurements had been taken by mechanical instruments. Advances in electronics during the war changed that. Following the war, when electronic instruments to measure details of flow and pressure were unavailable, they had to be created—either from scratch or by modifying existing instruments. The institute, led by

IIHR's post–World War II innovation in the switch from mechanical to electronic instrumentation was a matter of pride for Hunter Rouse (forefront), who here is showing off an electronic probe-moving mechanism used in the wind tunnel.

Phil Hubbard, did some of each. Pitot tubes (used for measuring flow velocity) were fitted with electronic transducers. Various pressure gages were modified and then built into models. Hunter Rouse developed a precision manometer for measuring gas pressure. When bridge-pier studies demanded the continuous measurement of sediment scour, a new device consisting of electrodes embedded in the pier fed the print-out. Perhaps the institute's major contributions were the hot-wire anemometer and hot-film anemometer, which (like Rouse's manometer) were developed and for many years manufactured here for laboratories elsewhere. These were capable of measuring rapid but small fluctuations in velocity in either air or water and thus were crucial for studies of turbulence.

The combination of these diverse efforts and the continued vigorous influx of government funding stood the institute in good stead. Its budget, research load, and staff size remained healthy and growing. Hunter Rouse reported that in the 15-year period from 1940 to 1955, the value of institute laboratory buildings had more than doubled, from around $100,000 to well over $200,000. Experimental and shop equipment, worth around $50,000 in 1940, had quadrupled in value. The institute's budget had soared from around $15,000 a year to an astounding $180,000, which even after accounting for inflation was a sixfold explosion—an increase that also characterized the number of full-time staff equivalents, which had soared from 5 before the war to 25 in 1945 and 30 in 1955, and which thereafter continued (along with the budget) a slow but steady increase. Staff members were often joined by a dozen or more part-time graduate students, who were now hired in unprecedented numbers to assist with contract experiments, an event that simultaneously offered them training in fundamental principles, research experience, and a salary. While select students had been offered this privilege before the war, thanks to support from the Graduate College or institute, the cooperative projects that had dominated the pre-war institute had imported their own staffs, and institute staff had been kept correspondingly small.

In the mid-1950s, the institute adopted a major new research initiative: ship hydrodynamics. Limited ship studies had been performed during the war, when the institute had tested the drag of ships in flowing water for the David Taylor Model Basin. There had been a ten-year hiatus in such efforts, but these were resumed with vigor in 1954 with the simultaneous arrival of one of David Taylor Model Basin's senior staff members, Louis Landweber, and the Navy-funded conversion of one of IIHR's basement river channels into a large ship towing tank, only the second inland towing tank in the nation. Landweber, who had been brought in to fill John McNown's vacated position, adeptly assumed responsibility for the naval projects. Within a year Rouse was commending him for "very capably [relieving] the writer (Rouse) of much administrative responsibility." Soon Landweber and his associates were carrying on a variety of ship-related studies, including analyses of wakes, the drag of truncated bodies, ship vibration, ship rolling, resistance from waves, and also continuations of turbulence and cavitation studies. Landweber did not confine his activities to research. He accepted the challenge of editing a translation of a Russian book on

wave resistance of ships and translating other professional documents himself. Landweber, trained in physics and mathematics, did
not think of himself as an engineer. His addition to the institute
strengthened Rouse's emphasis on fluid mechanics, as did the
arrival of another fluid mechanician, Enzo Macagno, in 1956. The
professorships of Macagno and Landweber became some of the
institute's lengthiest, with both remaining extremely active in research efforts into the 1990s, when they were in their eighties and
had long since formally retired.

The institute's research continued to diversify and, in the late
1950s and early 1960s, acknowledged an increasing number of
funders. The Navy continued to support efforts in turbulence, cavitation, instrumentation, and ship hydrodynamics, but also
diversified its interests, asking IIHR to look, for example, at flow
phenomena of jet-supported vehicles. The Rock Island Arsenal
funded a lengthy investigation of unsteady flow systems, such as
those in recoil mechanisms. Studies of flow in model river bends
were sponsored by the National Science Foundation. Stratified flow
and quasi-stable eddies were investigated with moneys from the
Army Research Office, and open channel resistance with funds from
the U.S. Geological Survey. Gate vibration was scrutinized through
a COE contract.

These were some of the projects that continued to propel forward IIHR's existence and efforts, along with the model studies and
applications-oriented projects that have traditionally supplemented IIHR efforts. For despite Rouse's insistence on fundamental
efforts, projects focusing on applications of hydraulics to problem
situations never completely ceased. Industrial contractors and government agencies continued to come to IIHR, seeking mechanics
skilled in constructing sophisticated models and researchers known
for their creative, practical solutions. Yet it would remain the task of
Rouse's successor to revive these industrial contract projects on a
major scale and to integrate them into the basics-oriented research
program that Rouse had so carefully molded.

The Russians' launching of Sputnik in 1958 had spurred a race
with the Soviets for scientific and technological leadership and resulted in increased funding for engineering education and leadership. However, the 1960s ushered in yet another shift in university
research funding, one that again reflected a change in the nation's

attitudes and values. Controversy about the Vietnam war spilled over into mistrust of the military and of government actions more generally. Many societal conventions were rejected while interest in the state of the natural world was reinvigorated. By the late 1960s, research support, particularly that for basic research, began to decline. Funding of applied research, especially that related to renewing the natural environment, took front stage along with legislation to protect the environment. At IIHR, the 1960s marked the first decline of ONR funding of IIHR research and equipment. Support dropped dramatically, along with ONR's domination. Facilities and equipment were no longer receiving the attention they required to maintain the institute in good stead. The strong merger of the trends of the times with Hunter Rouse's intense focus on pure research was unraveling. Once again the institute's financial security, as well as its course for the future, was questioned because of events outside its walls. Once again, it was time for a change.

Floyd Nagler had built the institute and established a premier research organization based on applied research and the use of small-scale models. His successors had kept it alive and kicking through the 1930s depression. Hunter Rouse and his coworkers had inherited a budget of several thousand dollars and built it into a budget of hundreds of thousands. In the process, he had shaped an internationally recognized institution with a research program based solidly on fluid mechanics. But it would require a different vision, one provided by Rouse's successor John F. Kennedy, to revive the institute one more time by broadening the base of the research program. Through combining problem-oriented, applied research with Rouse's search for underlying fundamental principles, and by pulling large-scale industrial moneys and hydraulic model work back into the institute, Kennedy and his institute colleagues would secure funding and projects that would add yet more new initiatives to the research agenda, refurbish the institute's infrastructure, and hoist the institute's annual budget into the millions.

8. The Science and Craft of Physical Modeling

Hydraulic models—proportionately accurate experimental miniatures of a real item, built on a scale small enough to be readily manipulated, observed, and comprehended—are, in large measure, the raison d'etre for the institute and its many annexes, which were constructed to house rivers, cities, ships, dams, and other hydraulic structures in miniature. If well-trained students and abundant publications have been the product of institute endeavors, then research with models has been the process through which they have been attained. While the institute is far more than a model shop, and while hydraulic modeling is but one approach to tackling complex fluid-flow problems, the quality and diversity of model work has become a defining aspect of the institute. Efforts with models of one sort or another have, from the start, dominated the institute in the funding they have brought in, the space they have occupied, the focus of researchers, and the number of theses generated. For these many reasons, the importance of models in institute research is worthy of explanation, as is the in-house shop that constructs these precision models and that through the years has become an integral component of IIHR's successful operation.

Historically, the use of small-scale models began to gain prominence late in the nineteenth century, when it became the impetus for proliferation of hydraulics laboratories first in Europe, then a few decades later in the United States. About that time, increasing human populations and industrial uses demanded more consumable water and better protection from excessive water during floods. These needs in turn required better understanding of flows in watercourses and around immersed bodies. Until then, most understanding of hydraulics had grown on a trial-and-error basis through field observations or through use of full-scale testing, both of which were very expensive. But as researchers came to better appreciate fluid mechanics processes and the laws of similitude theory, they realized the value and advantage of small-scale models. Laboratories then became a necessity, their space and structures accommodating the source of water, pumps to move it, gages to weigh or meter it, glass-walled tanks through which it could be observed and photographed, and other such apparatus.

The advantages of small-scale models became increasingly obvious. Projects using models were cheaper to complete and more easily carried out than field studies, and measurements could be made more precisely than in the field. With models, steady-state conditions could be established and variables controlled. Models thus allowed accurate assessment of the correct functioning of the complex river-training structures (such as massive dams and hydroelectric power plants) that were starting to be built. Because of the relatively small size, clean water, and features such as clear plastic walls of models, processes could be observed in the laboratory that seldom could be seen in the field. This ability to observe events directly led to the asking of new types of questions and the spawning of yet more thorough studies.

Models soon proved themselves capable of solving problems that could not be addressed either theoretically or numerically, and thus models have become an invaluable tool, a point illustrated by two IIHR projects in the 1970s. The first involved an examination of ocean currents and waves in the vicinity of a water-intake structure for a Florida nuclear-power plant. This plant draws in water for its operation through massive 12- and 16-foot-diameter pipes that extend a quarter mile out into the ocean. There the pipes connect to an intake fitted with protective concrete slabs that prevent sea life from being sucked into the plant. One day a diver swimming near the intakes found himself unable to avoid the ever-tightening eddies that pulled him downward into a blackened tube. Swept onward without recourse, he yielded to the flow. Ten minutes later, he popped up in a lighted chamber and found himself in the cooling water intake area of a nuclear-power plant, where he was promptly arrested for sneaking into a protected structure. Later investigations showed that the concrete caps sheltering the pipes had fatigued because of

Investigations that use small-scale models have always dominated IIHR's research. Take as an example studies of local scour of riverbeds around bridge piers (top), which became a major research focus in the 1950s. The scouring process could be studied more cheaply, easily, and precisely in the laboratory setting with proportionately accurate miniatures of bridge piers (bottom) than it could in the field.

the constantly reversing pressures of the pounding waves. The pressures were far too dynamic and complex to describe theoretically or numerically; a hydraulic-model simulation of the ocean floor and inlet pipes under varying states of the sea was clearly required. IIHR thus constructed a model of the inlet cap and tested it under varying wave conditions. An inlet cap was designed that not only would exclude wildlife and divers, but also would not collapse as its precursors had done.

In a second example, an electric power utility was troubled by sediment clogging the water intakes of its hydroelectric power plant. IIHR performed a model study costing approximately $50,000. By developing structural measures that eliminated sediment deposition, IIHR eliminated annual dredging costs of hundreds of thousands of dollars a year. Similar design improvements and efficiencies developed through hydraulic modeling commonly save clients a great deal of money. Director John Kennedy used to quip that IIHR could survive handsomely if it received a mere 5% of the money that clients saved because of model studies performed here.

For all of these reasons, the use of models became an accepted procedure in designing hydraulic structures. Models also proved their use in questions of maintenance and repair. With models, expensive and perhaps irreversible modifications, upkeep procedures, and repair practices can be investigated before they are applied to structures costing millions of dollars. In 1996, for example, IIHR used a hydraulic model to check the consequences of an accidental injection of foam into a concrete tunnel that is part of a nuclear power plant's emergency service water system (ESWS). The foam had been applied in an attempt to seal seeping cracks that had formed in the tunnel. However, large quantities of foam had leaked into the tunnel itself, potentially reducing the amount of water that could flow through the ESWS. Divers were hired to remove the problem foam from the tunnel. IIHR was then asked to determine the seriousness of the threat. Obviously, repeated experimental foam injections could not be made into the tunnel itself. Instead, a hydraulic model of the tunnel was constructed, and foam pieces were released to determine in retrospect how the foam had moved through the actual power plant's tunnel.

While models have always been a prominent component of

IIHR's research, their use has evolved through the years. The very first experimental work performed here involved the use of models to design a proposed dam on the Cedar River south of Cedar Rapids. Following that project, the Hydraulics Laboratory started modeling the components of dams, spillways, and locks, in addition to straightened, bent, and flooding river reaches. Model tests performed in 1929, concerning the Des Moines River, were reputed to be the first model studies performed in this country to determine, in advance of field construction, the advantages of straightening a river. Modeling also was used to investigate aspects of flow fundamentals as well as fluid-transport processes like alluvial-sediment transport. The expansion of the Hydraulics Laboratory in 1928 and again in 1932 provided space for models to continue increasing in size and number.

Some of these model studies, such as those involving the dam and power plant at Keokuk, Iowa, have gained a historic significance of their own. That dam's massive structure, with its 30 turbines and 119 spillway gates, made it the world's largest water-power plant in 1913, the year of its completion. The amount of concrete used in its construction was exceeded only by the amount used for the Panama Canal. Floyd Nagler was asked in the 1920s to determine the Mississippi River's flow over the dam's spillways, which he accomplished with both field studies and laboratory models. The institute worked with these spillways again in 1954, when it performed tests for maintenance operations on the spillways and their foundations. By 1996, design adjustments were required again. Water's constant pounding had by this time eroded 18 inches of spillway concrete in some spots, challenging the dam's integrity and safety and slowing water passage over the spillways. Techniques were needed for high-quality but economical repair. IIHR was again contacted for help—in part because of the historic relationship between the institute and the dam.

Nagler's 1924 spillway-discharge tests for the Keokuk Dam were influential in convincing American engineers that a model could indeed be used to predict reliably a prototype structure's behavior. This project helped stimulate a flood of model studies at IIHR and elsewhere. During the 1930s, IIHR served as the major site for COE model work on the lock-and-dam facilities that reshaped

In the 1930s, IIHR's models of the upper Mississippi River's lock and dam system (such as this erodible bed model of the Alton, Illinois, area) prompted a newspaper journalist to describe institute researchers as "an army of Gullivers astride miniature rivers, locks, bridges, trestles and dams."

the Upper Mississippi and other Midwestern river systems. The following decade, IIHR contributed to war-related efforts by modeling gas diffusion over cities, wind movement past simulated mountains, water movement past torpedo heads, and ship hydrodynamics.

Hunter Rouse, during his tenure as institute director, forged his own stance on model studies. A historic summary written during his directorship acknowledged that "essentially all experimental research in hydraulics proceeds by means of models, for the subject matter is far too complex to permit many flow principles to be studied without relation to a specific boundary geometry." However, for the most part, Rouse's models consisted of precise geometric or idealized shapes used to investigate fundamental principles of features such as boundary layers, cavitation, and fluid flow. Although applied hydraulic-modeling studies of specific dams, rivers, spillways, etc. continued to some degree, Rouse actively endorsed such studies only if they promoted an understanding of fundamentals.

"It is true that experimental hydraulics in the past was greatly furthered by the advent of model testing," he wrote. "Those lessons have long since been learned, however, and continued progress will not come from repeating events of the past. . . . How much better if a fraction of the cost of the specific tests were put into a general study, thereby reducing expenditures for future tests many-fold!" he exclaimed when considering the routine testing of first one model, then another, in order to discover how to construct a given dam or steer water over a certain spillway. Such tests, he thought, were better performed in testing laboratories, not in a university institute such as IIHR, which Rouse was attempting to dedicate to the exploration of fundamental principles and theoretical fluid mechanics. "It is the Institute policy to undertake model studies of a specific nature only if new principles of sufficient interest to warrant their investigation are involved," he wrote in another article.

In contrast, Rouse's successor John Kennedy adamantly believed that applied and basic research nourished each other, with applied research (including hydraulic modeling) revealing the problems that stimulated basic research and in turn feeding off the knowledge gained by basic research. Applied research also attracted funds that could be used for broad research programs.

During the Kennedy years, four new annexes were acquired and occupied by models and test equipment, and a fifth was built in 1996. This space and equipment stimulated and redirected research efforts. The opening of a low-temperature flow facility in 1970 engaged IIHR in the new field of ice modeling. The construction of a large closed-circuit, low-turbulence wind tunnel and smaller portable icing wind tunnel, many straight and meandering flumes and basins, and a large environmental flume provided additional opportunities. Modeling equipment was supplemented in 1967 by the institute's first computer, which had a dramatic effect on all research projects and initiated a major expansion of research capabilities. The computer, in tandem with lasers and other steadily improving measuring devices, greatly speeded up and improved the accuracy of automated collection and reduction of model data, so that many more variables could be recorded simultaneously over a much larger area. The versatility and number of model projects shot up in response.

The Kennedy years were marked by a responsiveness to burgeoning environmental concerns. Large model studies were initiated to investigate cooling-tower systems for power plants, thermal effluent from cooling water–discharge systems, and dam passageways for migrating salmon. Such projects demonstrate the applicability of hydraulic research to society and everyday life. But then, hydraulic-model studies have always brought home the significance of IIHR's research. Institute models have helped medical researchers understand the dynamics of fluid flow in the urinary tract and peristalsis in the small intestine. Transparent models of plumbing systems have determined safe construction set-ups and demonstrated the dangers of water pollution through back-siphonage. Designing more efficient pump intakes for power plants has improved the performance of power plants, thus holding down the cost of electricity.

In a lighter vein, IIHR's model research has illuminated the dynamics of flow around fishing lures, outboard motor propellers, and racing swimmers. IIHR's interest in the propulsion of swimmers, conducted jointly with the Department of Exercise Science, began in the days of Hunter Rouse and has continued intermittently through the century. Members of Iowa's swim team have improved their techniques using videotapes showing them executing various strokes in the laboratory's large flume, or swimming against various speeds of water flowing through the flume. The dynamics of flow around their bodies have been studied by watching the movement of dyes, pellets, air bubbles, and various kinds of tufts glued to their bodies. In the 1980s, models of the human hand were embedded with pressure cells, attached to rods, and swept through the water, and cylinders simulating legs were gyrated in a kicking pattern, all in an attempt to better understand the flow of water around the complex shapes of the human body.

Our lives continue to be touched in an amazing diversity of ways by IIHR's multifaceted model research, which has helped assure that mussel beds will remain unsilted, that waste water will drain efficiently away from our cities, and even that our passage over suspension bridges will remain safe. In 1940, the Tacoma Narrows bridge, affectionately referred to as Galloping Gertie, arched gracefully above Tacoma, Washington, its 2800-foot span forming one of the world's longest bridges. The suspension bridge's slender, sinuous lines, praised for their grace, were the culmination of a widely accepted trend to lighten and streamline bridges. Thus, no one worried when the wind swayed the relatively thin, light bridge so greatly that even practiced construction workers became nauseous. Once open, the "galloping" bridge attracted motorists who were awed at the sight of cars in front of them almost disappearing from sight along the undulating surface that corkscrewed like a twisting ribbon. However, on a stormy day after only four months in operation, Galloping Gertie started to oscillate wildly, swaying into semi-vertical positions, and a few hours later Gertie collapsed into the water in one of the most notorious bridge failures of all time. The reasons for this collapse were extensively investigated, with the bridge engineer David Steinman playing a major role in the heated debates that followed. Rouse was widely enough respected to be drawn into these debates by Steinman, who retained Rouse to use the institute's wind tunnel to determine the aerodynamic pressure distribution along model bridge segments. Steinman incorporated these model tests into a presentation that he made on the aerodynamic stability of bridges (among other structures) at the institute's Third Hydraulics Conference in 1946, and one of Rouse's graduate students in 1948 wrote a thesis that further discussed the susceptibility of various model bridge deck and girder systems to destruction by wind.

Successful model studies are not simple affairs. Each model is a unique custom product, requiring skilled shop personnel who are creative, knowledgeable, committed to excellence, and dedicated to working constructively with the institute's researchers. The design process itself is complicated, involving simulation of an exact set of features that are both structurally and proportionately accurate. Design decisions regarding model scale, data required, instrumentation, and the like must be melded with practical considerations of cost, facility limitations, and instrument capabilities. Construction must be completed with precision and care, since a small distortion in the model is greatly magnified when scaled up to real size. And seldom does a model operate as it should the first time around. It usually has to be calibrated. Models typically are extensively rebuilt until trial and error mix with design to create success.

As early as the mid-1930s, IIHR's shop had become skilled at meeting these many challenges. Assertions about its capabilities were then used as a selling point to attract government moneys to the institute. Mavis, when writing to the COE, boasted that "the laboratory at Iowa City is furnished with a very complete shop for the construction of wood and metal models." He pointed out that Iowa's shop was located in the laboratory (rather than at a "considerable distance from the laboratory"), and that in Iowa, shop "personnel of all grades have been carefully organized and trained over a long period" and were skilled in locating the numerous types of materials needed for the models. The fact that IIHR's shop has always been located on site and has hired its own staff has been a major part of the institute's success with models and has distinguished it from other hydraulics laboratories.

In 1997, eight permanent employees worked in the mechanical shop, with additional carpenters and laborers hired for periods of several months when labor-intensive models were being built. In addition, IIHR shop personnel have traditionally been long-term employees who have developed an understanding of hydraulic models and close relationships with the institute's directors and researchers. When a tornado screamed past Rouse's home in the 1950s, shop personnel harvested the large walnut trees that had fallen on his property, cut and dried the lumber, and fashioned it into bookcases for his office. Former shop manager Dale Harris spoke fondly of how he grew with the laboratory, coming here in 1940 as a lab technician straight out of high school and working first under Rouse and then under Kennedy. Employee Stanley Stutzman started here in 1960 and retired 38 years later. Such work records not only magnify experience and expertise; long-term experienced shop personnel become involved in the design of models, interpreting drawings and offering practical solutions to thorny problems. They also develop a sense of community that encourages the exchange of ideas and deepens a person's commitment to the lab's success.

There's a sense of pride, conscientious effort, and high expectation that has typified shop operations from the start. Nagler built a laboratory so solid that later remodeling efforts were hampered by the thickness of the walls and floors. Rouse was known for not doing anything unless he did it well. Just as the institute's directors have demanded much of themselves, so they have set the tone for others. IIHR's shop has gained a reputation for the quality and pre-

The high quality and output of IIHR's shop has always depended on the skills and dedication of employees such as former shop manager Dale Harris (forefront), who worked here for 44 continuous years.

cision of its efforts, with fine workmanship and tight tolerances being a tradition. "If we've built a model that's not up to snuff," boasts shop manager Jim Goss, "I've seen the crew rip it apart and rebuild it from scratch on their own time, just to get it right." Frank Turner's fiberglass models are another example of this craftsmanship. Turner is an old-school pattern maker who forms by hand the molds for very complex shapes that in industry would be cut using computerized directions. Such manufacturing would not be cost-effective for the lab's one-of-a-kind models. So instead Turner, using skills that are rapidly being lost from the workforce, crafts the molds that then are used to cast model ship hulls, airfoils, turbine intakes for hydroelectric plants, and other complicated bodies in miniature.

The institute is unusual in the neatness of its efforts, even during the construction phase. Hydraulics laboratories in general are because of their efforts bustling, wet, noisy places, littered with models in various stages of construction, revamping, and disassembly.

But "until I visited other labs, I never realized how unusually well-ordered and how spick-and-span our lab was," says Phil Hubbard, who earned a doctorate here in 1954 and then joined the faculty. "There were no pipes suspended at head level to duck under, no peeling paint, the floors were clean, no dripping water or puddles like there were in labs I later visited elsewhere."

Speed and drive are other components of the formula. With the study of the nuclear power-plant cooling tunnels mentioned earlier, for example, eight shop personnel working overtime were able to complete the model within a week. The following week, the company's engineer visited to observe the initial model tests, and a half dozen graduate students were pulled in to complete the tests within a few days. "When a client is coming the next Monday to see a model in operation, we guarantee that it will be up and running by that day," states Goss. The sense of assumed quality carries through a project's report stage, so that clients assume that the results of careful work will be transmitted with pride.

The shop's performance results in contracting agencies expressing a high degree of satisfaction. Clients often come to visit the laboratory when their models are operating. These visits serve educational and practical functions. In a model, where flow is easily seen because of the model's smaller scale, clean water, and transparent structures, clients can finally perceive and understand processes that until then may have stumped them. Squirt a filament of dye into the water of a model dam, and suddenly a fisheries biologist can comprehend the convoluted passage of salmon through a dam diversion structure. Hydraulics consultants and design engineers standing around such a model find the process equally stimulating. Soon they are brainstorming with IIHR personnel about alterations that would improve a given structure's performance—a new shape for an outfall pipe, for example, that will prevent young salmon from being plunged to the bottom of a river into the waiting mouths of predators.

These novel ideas might never have been conceived had the clients not seen the model in operation. When new approaches can be investigated using the model at hand, shop personnel have been known to work around the clock to alter a model for testing in the morning, while clients are still in town. "We found that the fish were not letting themselves be swept downstream through the outfall,"

Larry Weber explains about a salmon-diversion project completed in 1994. "We realized that the fish were fighting their passage by laying low in the sluggish corners of square pipes, which had been chosen for ease of construction. But a round pipe would be needed to equalize water velocity and pass the fish through more efficiently. We asked the shop staff to make the change, and by the next morning we were testing a successful passage structure. The clients were delighted at both the speed of the transformation and at the effectiveness of the final model structure."

While shop personnel deserve much credit for the success of IIHR's model studies, the shop does far more than build and maintain models. It retains personnel and instruments to collect field data. It constructs and maintains instructional and other experimental equipment, including the structures that house models and experiments—the various flumes and conduits through which watercourses and the wind tunnels that hold and direct flowing air. Innovative equipment has been designed here and constructed to unusually fine tolerances. In the large recirculating wind tunnel, for example, that moves air at speeds up to a hundred miles per hour, careful design and construction have produced extremely smooth flow that has eliminated problems with interference from unwanted air turbulence. Items have frequently been designed here for university research laboratories elsewhere that cannot produce such equipment for themselves. Shop personnel have constructed custom-designed wind tunnels and flumes of various sizes and capacities, loaded these onto trucks, traveled with them to other institutions, and got the equipment installed and operating. These items are crucially important to hydraulic modeling and experimentation but are sometimes highly specialized and not commercially available.

IIHR and its shop also have a history of designing, testing, and manufacturing electronic instruments. In the early 1950s, Phil Hubbard developed a hot-wire anemometer here and started manufacturing and selling it through his Hubbard Instrument Company, which operated through the institute. In later years, he fabricated a variety of electronic instruments for specific research projects that did everything from measuring the scour around bridge piers to the blood flow in humans. Meanwhile, Rouse had developed a precision manometer, which accurately measures

Many electronic instruments have been developed or modified at IIHR, and some of these have been sold regularly to other institutions. Here Phil Hubbard lectures graduate students on the time constant for the thermal lag and on the functioning of his hot-wire anemometer, a device that measures fluid velocity by using wires one-seventh the width of a human hair to assess the rate of heat dissipation.

small differential air and fluid pressures; it has been manufactured and distributed by IIHR ever since. Among other instruments that are manufactured and sold here are point gages (that measure water-surface elevations and bed profiles), standard two-tube differential manometers (for measuring pressure differences between two points in a flow), pitot tubes, and in past years mercury gages and air-jet tables. Such instruments are constructed during research down-periods and help maintain a steady workload in spite of the fluctuating demands of research projects. Their production and sale bring funds into the institute and, equally important, provide a crucial service to researchers, educators, and institutions elsewhere.

The use of models at IIHR continues to evolve, as do the instruments and equipment associated with model studies. Where students once laboriously collected data with physical measuring devices and a notebook in hand, in the 1990s they employ remote recording devices that automatically feed data points directly into a computer. Instruments continue to evolve so that they can handle greater numbers of data points simultaneously, and they do so increasingly without using probes that extend into the air or water and may disrupt the flow. IIHR has invested heavily in emerging high-technology data-acquisition systems, such as the newer breeds of laser-based velocimeters that nonintrusively record multiple changing water parameters on a computer that instantaneously analyzes the results. Systems such as laser Doppler velocimetry (LDV) and acoustic Doppler velocimetry (ADV), for example, allow one-, two-, and three-dimensional noninvasive velocity measurements to be taken up to hundreds of times per second.

All of these improvements enable researchers to ask questions that could not even be imagined before. In fact, the two feed each other. Complex instrumentation pushes researchers to investigate new problems, and new research programs stimulate equipment development. This resulting growth of research has been obvious in the physical growth of the institute and in the diversification of the shop and institute support units over the past 75 years. For more than half of the institute's existence, shop personnel consisted solely of mechanics, each of whom had his own specialty (such as welding, machining, or woodworking). Experts in electronics started to emerge following World War II. The first such expert was a faculty member, Phil Hubbard, whose electronic innovations helped provide state-of-the-art instrumentation to IIHR's researchers. Shop personnel assisted with instrumentation when necessary. Hubbard left the institute to become the UI's dean of Academic Affairs in 1966, the year after Jack Glover joined the faculty. Glover also directed the fabrication of electronic instruments used in experiments here and, equally important, oversaw the installation of the first on-site institute computer in 1967.

By the early 1970s, IIHR's shop included an electronics unit that was distinct from the mechanical shop and was occupied by a full-time electronics technician. Finally, when Glover left the institute in 1980, a decision was made to replace him not with a faculty person but with a professional engineer who had expertise in computer and electronics systems. Thus, in 1981 electronics and instrumentation functions were placed under the supervision of a new staff member, Jim Cramer. Cramer's job was rapidly reshaped by the proliferation of personal computers at the lab, and another engineer, Ken Hartman, was hired to assume the electronics and instrumentation activities so that Cramer could devote his efforts totally to computer systems. In 1997, the institute boasted three separate support units: a mechanical shop headed by Jim Goss, an electronics unit manned by Doug Houser (who replaced Hartman in 1986), and a computer support unit led by Mark Wilson.

Computers are thrusting hydraulic research into new realms and capabilities, just as hydraulic modeling did nearly a century ago. Computers can analyze situations such as ship propeller-hull interactions or turbulence with far more detail and sophistication than was ever possible before. However, some situations are still so poorly understood or complex that, for the immediate future, tinkering with physical laboratory models remains the only way to grasp them. And just as field studies have been used to verify the results of hydraulic models, so hydraulic models will remain necessary to validate the results of computer models. Therefore, IIHR has chosen to invest resources in maintaining sophisticated experimental facilities as well as supercomputer facilities. Hydraulic models and the shop personnel who produce them remain as crucially important as they did when the Hydraulics Laboratory consisted of little more than a small brick cubicle. The Hydraulics Laboratory was founded on the belief that well-equipped, skilled shop personnel producing quality models and experimental equipment were crucial to the laboratory's success. Seventy-five years later, that belief remains firm.

9. The War Years

On November 21, 1945, Hunter Rouse wrote the following to Dean Dawson of Iowa's College of Engineering:

> In accordance with your recent request, I am submitting herewith a brief report on the activities of the Institute during the past several years. . . . No formal report of this nature was submitted during the war, due in part to the pressure under which we were working and in part to the fact that practically all of our output was of a classified nature . . . .

Rouse, true to form, was being direct, succinct, and accurate. The war years had indeed been busy, with researchers and shop personnel working late into the night to obtain urgently needed results, then traipsing home for a few hours of shut-eye, only to return to the lab at 8:00 the next morning to continue their efforts. The tremendous volume of defense-related projects had demanded such longstanding efforts. Graduate students quit their studies to join the staff full-time, their salaries being paid by contracts from any of the several war agencies that had approached IIHR for help, most commonly the National Defense Research Committee (NDRC).

Long hours, at least for Rouse and his shop manager Dale Harris, were also demanded by the need for secrecy. At night, when others had deserted the building, these two could talk and complete their confidential tasks without fear of prying eyes or ears. During the workday, construction of a classified experimental model might be partitioned among several shop workers, each one assembling a model section but no single worker being shown the entire plan or told the complete story. Then, as the daylight waned and the lab emptied, Rouse and Harris would assemble the sections to form a single experimental setup and commence testing. As midnight approached, Harris and Rouse would walk a block north to a tiny diner perched on the bank of the Iowa River and there renew themselves with ham and eggs. Then back to the lab again until two or three in the morning, when the absolute need for sleep finally pressured them to turn off the pumps, return home, and collapse into bed.

Occasionally, laboratory workers were forced to expose their efforts to the eyes of Iowa City residents—to shoot flamethrowers into the early-morning fog, for example, firing them from the Burlington Street bridge down the Iowa River. The research was classified. Questioning passers-by could not be told the truth. "We're testing apple-orchard warmers," they would be told, or "We're developing ways to warm the air over orchards to prevent frost damage."

In reality, the "orchard warmers" were hopeful mechanisms for dispersing the thick fog from British landing fields. The institute was secretly helping to perfect FIDO, the "Fog, Intensive, Dispersal Of" system intended to bring Allied aircraft safely back to landing fields after completing bombing raids over Germany. The thick, soupy fog that at times blankets England was causing heavy losses among such aircraft, and radar—only recently developed—had not yet been applied to guiding the landing of fighters. Thus, the British had experimented with various fog dispersal mechanisms. They mounted large blowers on trucks and hauled them around the runways, trying to blow away and evaporate the moisture with heated air; the trucks got stuck in the mud. They lined the runways with trenches and tried burning first coke, then fuel oil, to heat the air and evaporate the moisture; the coke warmed the air too slowly, the fuel oil emitted a thick black smoke that blanketed the airfields more heavily than the fog itself. The British requested help from U.S. engineers, and IIHR started investigating other air-warming mechanisms, developing and testing the burners that resulted in a successful FIDO system: one that consumed high-octane gasoline in interconnected burners lining the side of a landing strip. The system was costly. On the average, 30,000 gallons of the scarce fuel were used to bring in a single airplane safely. But the system worked well, and the cost was deemed acceptable. Gasoline lighted quickly, heated the air sufficiently, and produced a large patch of clear air above the landing strip. Rouse explained this all to a group of lunching Rotarians shortly after the war, when he verbally retraced the flight of a bomber back home:

> The fog was thick as pea soup but ahead of them, the fliers could see a faint yellow glow. They were guided by radio toward this steadily increasing patch of yellow in the fog. Suddenly a transparent puddle opened in the fog below the plane, and the men could see their home field clearly. The pilot said that that was the most pleasant thing he had ever seen.

Not all war efforts produced such glorious effects. That's not to say that the more mundane war projects were not important or needed. "Our army does all right with its interceptor planes, but it was out-flanked by a lot of sewage when it tried to find adequate and efficient grease interceptors to install in the great camp kitchens," a newspaper reported when the institute expanded its prewar work with grease traps to test and certify mega-traps for army kitchens. And as commonplace as plastic plumbing pipes may be today, a newly developed tubing made from Saran needed to be tested to determine whether it could be satisfactorily substituted for copper tubing in war plants and emergency housing. (The institute confirmed that it was fine for cold water pipes, but that the plastic softened when the water became too hot.)

Fire-fighting nozzles might at first seem similarly mundane, at least until one considers the image of a troop-laden ship burning to ashes without recourse. Quenching a fire on such a vessel depended on powerful equipment capable of accurately directing a focused arch of water toward the blaze. A Coast Guard officer fiddling with nozzles had found that placing a vaned barrel directly in front of the standard nozzle increased the range of shooting water by about a third. The information was sped off to Washington, and soon IIHR had signed a contract with the Coast Guard and David Taylor Model Basin to investigate the existing forms of monitors (that propelled the pressurized water) and nozzles, and to design improved forms of both. Testing soon commenced in the basement of the Hydraulics Laboratory and outside, over the Iowa River. The stream of water shot from the existing, inefficient Coast Guard monitors predictably and rapidly disintegrated into a cloud of spray. However,

"A humble but necessary item is the grease interceptor," wrote a journalist about IIHR's efforts to develop large grease traps for U.S. Army kitchens. Research involved pouring grease into funnels above potential interceptors and determining how much escaped into adjacent holding tanks.

contrary to the commonplace notion that the broadening of spray was induced by the resistance of surrounding air, testing showed that the water jet was already turbulent when it left the monitor, the disintegration having been set in motion by eddies established within the monitor, roughness in the nozzle's walls, and certain nozzle shapes. The resulting turbulent jet could actually fuel a flame by pumping significant quantities of air into the blaze. IIHR's subsequent modifications of nozzles and monitors were directed toward decreasing the sources of such turbulence. Research efforts eventually produced a nozzle that increased the amount of water concen-

Coast Guard fire-fighting nozzles delivered insufficient amounts of water to burning ships because they produced a dispersed, air-laden spray. IIHR tested and developed monitors and nozzles that delivered a more focused, powerful water stream.

trated on a target by as much as 900% at a 90-foot distance from the monitor, with the water jet itself traveling as much as twice as far as it had with the less efficient monitors.

In order to test the water jets, the laboratory's shop personnel had remodeled the basement of the institute, converting one of the lengthy flumes into an enclosed test gallery. They had fitted the gallery with eight sizable fire pumps at one end, a recording target at the other, and equipment mounted on a traveling carriage to sample the concentration of the water jet along the way. This was but one of the major laboratory refurbishings forced by wartime research. Others involved construction of the institute's first wind and water tunnels.

Water tunnels—closed circuits within which a steady flow of water is pumped past an object under scrutiny—differ from open-surfaced flumes by a single major variable: while the speed of water is adjustable in both, water can be pressurized to varying degrees only within the closed tunnel. Thus, tunnels are required for studies of cavitation—the problematic boiling of water in low-pressure areas created when objects speed through water. Studies of cavitation are crucial to the development and design of fast-moving underwater bodies used in naval warfare, such as torpedoes. Water tunnels constructed specifically to test ship propellers had first been developed around 1900. World War II created a great need for a more general variable-pressure water tunnel for testing flow around bodies with or without cavitation, one that could provide crucial war-related underwater ballistics tests. The institute became one of two nongovernmental bodies to possess one of these relatively new tunnels, the other being the California Institute of Technology.

The institute's initial, small tunnel was built in 1941 to teach about cavitation in its upcoming National Defense Training Program class. Two years later, it built a larger unit, fitting sections with tempered glass for observation purposes. Objects of various shapes could be mounted behind this glass, water pressure and speed could be adjusted, and the objects could "thus be subjected to practically any desired degree of cavitation." These experiments commenced an NDRC-funded steady series of tests of bodies with variously shaped heads—conical, ellipsoidal, rounded, blunt, and others. The pressure distribution was measured and vapor pocket photographed at various degrees of cavitation, all in an attempt to provide the heretofore nonexistent data necessary for the design of nose shapes of underwater projectiles such as torpedoes.

The year 1943 seemed to mark an expansion of general research as well as war-related efforts, in part because of the installation of IIHR's first two low-velocity wind tunnels. One was a small, portable tunnel, built purely for visual and photographic investigations of fluid flows (although use was soon shunted into experimental applications). The other was a massive affair occupying the entire north end of the second floor, including its exterior structure: the building's four north windows were removed and the resulting holes encased so that the air flowed into the rectangular 6-by-4-foot tunnel, discharging through windows on the east and west sides of the building (which also were removed for this purpose.) A large fan was installed to draw the air through the tunnel from one end to the other, with air speed being controlled by fan speed and encased louvers.

Construction of the large wind tunnel had been funded by NDRC "in order to permit certain aspects of chemical warfare to be explored rapidly at reduced scale." More specifically, the Armed Services had wanted to determine whether gas-diffusion tests in the wind tunnel could ascertain the correct drop-position and dose for chemical-warfare bombs. Instruments were mounted on a moving carriage and tests were commenced on how best to inject smoke (a proxy for the chemical gases) into the chamber and then measure its path. The wind tunnel proved its worth: while there were problems with the rapidity with which instruments measured the changes in gas concentrations, "the results of such tests in the wind tunnel are made directly applicable to field conditions," and tests performed within them would "permit computation of the rate of gas release required at any wind speed to produce lethal concentrations over any desired zone." However, those concentrations might need to be larger than anyone had previously thought, for wind-tunnel tests showed that most of the gas would spread upward rather than outward through a town.

Preliminary chemical-warfare tests completed, use of the wind tunnel was shunted to urgent efforts with the FIDO systems described earlier. Then suddenly in 1945, priorities governing use of the wind tunnel shifted once again, and gas-diffusion tests recommenced. On June 8, 1945, exactly a month after the end of the European war and two months prior to the dropping of the first atomic bombs on Japan, the institute released a confidential report directly to the War Plans and Theater Division of the Chemical Warfare Service (CWS). This report responded to the charge "to obtain data indicating the variation of gas concentration with position in a typical Japanese urban district when gas is released from standard CWS gas bombs." The research techniques developed two years earlier were employed once again, revealing that a large portion of the smoke was found to sweep up and over the tiny model buildings rather than to spread laterally to locations where people would be found. This report also stated clearly the need for correlation of model studies with field tests. Had the results of this report been more positive, who can guess the effect it might have had on the closing days of World War II?

Additional NDRC-funded wartime projects sought to determine whether measurements of wind currents over models of various

Shortly before the end of the war, the wind tunnel was employed to test diffusion of gases over a 1:72 scale model of a typical Japanese residential-manufacturing district constructed in the wind tunnel.

mountainous landscapes would yield useful data on weather phenomena. These measurements of atmospheric diffusion over mountainous terrain were performed in the wind tunnel. And early in the war, a confidential photographic study of ocean-wave movement had been conducted, and a defense-related investigation of stratified flows and turbulence focused on the characteristics of mixing at the ocean thermocline. This study was significant because it was the institute's initial research on flow of fluids with density stratification, an area of investigation that would be continued in post-war decades.

Another trendsetter concerned the drag of stationary ships in flowing water. The majority of ship-model tests involved towing the models through a long, narrow water channel, appropriately named a towing tank. However, the Navy's ship-modeling laboratory, the David Taylor Model Basin, decided during the war to attempt the reverse by using flowing water and stationary ship models—similar to the studies of stationary aircraft in wind tunnels. While the large recirculating channel was under construction at the Model Basin, the institute performed preliminary studies on a more modest scale. They were intended to preview the types of problems that might be encountered with the Model Basin channel and to explore methods for controlling turbulence in recirculating flumes. These particular explorations commenced in 1942 and continued only for a few years. However, they were significant in being the first institute ship-model tests, representing the commencement of government-sponsored ship-hydrodynamics studies that were to become a major thrust in the second half of the century.

In these many ways, IIHR supported the Allied forces on the battlefields of World War II, as did fellow hydraulicians across the country. As Gail Hathaway, a leading hydraulic engineer in the U.S. Army Corps of Engineers (COE), wrote, "In the mobilization of total resources for prosecution of war and hastening of victory, the hydraulic engineer played an important part. All of the acquired experience and knowledge of engineers in various phases of the hydraulic field were utilized." Hydraulic engineers designed the artificial harbors and portable breakwaters that followed the Allied troops onto Normandy's beaches, there supplying soldiers as they thrust forward through France to Germany. When the troops reached the Rhine, hydraulic engineers estimated the river's characteristics, so that pontoon-bridging and amphibious-equipment operations could safely move armed units across into Germany. What if dams were to be considered an instrument of war, their explosion producing floods that could destroy as efficiently as a bomb? Hydraulic engineers estimated the size and duration of possible floods and recommended positioning troop camps accordingly.

And then, finally, the war ended. The urgent demands for immediate research results ceased. No longer did Rouse and Harris trek to the corner diner at midnight in an effort to stave off fatigue. No longer were stories of orchard warmers fabricated to camouflage the true reason for confidential tests. Researchers once again returned home to their families in the evening. Students turned staff members returned to classes on a part-time basis, many bringing along thesis topics spawned by their war efforts.

Rouse tallied up the costs of the institute's contributions to victory. "In view of the full attention which has been devoted to war projects, it is not surprising that staff publications during these years have been limited to six papers, ten discussions, and one book," he wrote to Dean Dawson. However, he showed no signs of begrudging the war's demands. Already hydraulic engineers and hydrologists across the country were turning their thoughts to peacetime uses of the many discoveries that had been propelled by the war's urgency. Radar, touted as one of the most spectacular developments of the war, was first used for thunderstorm detection in 1943. Revolutionary techniques such as the use of echo-sounding devices were applied to surveying the depth of shallow waters. Air photography proved useful in the study of underwater structures. Rouse resumed plans for the hydraulics conferences that had been interrupted by the war, and in 1946, the institute hosted its third such conference, focusing on the topic of post-war applications of war research.

Rouse seemed to sense that the institute would never return completely to what it had been. The war had proven itself to be a watershed in numerous ways, bringing new staff, equipment, government funding, and initiatives, all of which had redirected IIHR's research away from its prior applied efforts and toward fundamental fluid-mechanics investigations. The institute's new course set during the war would be maintained through the coming decades. IIHR's staff, expanded to 25 active researchers exclusive of personnel in the USGS and COE offices, would not decline. Water and wind tunnels would continue to play an active role in research. In fact, Rouse became a staunch advocate for the construction of air tunnels, explaining that "staff members of the Iowa Institute were initiated perforce" into their use by the war "with results which were pleasantly surprising," and that he now regarded them as an essential piece of equipment for a hydraulics laboratory. While research on sediment, plumbing, and other pre-war topics was resumed, these foci were joined by others that had been initiated during the war. Studies of the jets and nozzles of fire monitors continued for only a few years beyond the war, but the FIDO studies spawned more fundamental post-war research in the two- and three-dimensional flow conditions of submerged jets. Other war-initiated research—in particular ship hydrodynamics, cavitation, and flows of fluids with density stratification—would remain a significant emphasis for decades, with ship studies becoming a creative and financial mainstay throughout the century.

The new fields were funded with research alliances that had been sparked during the war. Federal defense departments had learned the efficiency of contracting with university research laboratories to provide basic information and services. Now several of the many NDRC-funded projects were picked up by other federal agencies—the investigations of wind flow over mountainous terrain, for example, by the Air Force. The biggest new player was the Navy's Bureau of Ships which, in March 1945, contracted directly with the institute for an extensive program of research, including the peacetime continuation of World War II cavitation studies in the water tunnel, drag studies of stationary ship models in flowing water, wind- and water-tunnel studies of turbulence and its control, and fire-monitor and nozzle studies. The contract was written in a very general manner so that "a wide variety of basic problems in hydrodynamics may be investigated as the need arises," as Rouse wrote to Dean Dawson. In retrospect, if the evolution of institute research is considered solely in terms of its principal funding agency, then the COE's pre-war support of research was replaced by NDRC during the war, which gave way to the Bureau of Ships once hostilities ceased.

Soon after the war, Rouse seemed to mimic his predecessor Floyd Nagler, who, immediately upon completion of one major laboratory expansion, would start politicking for another. "It appears likely that the program of basic investigations for the Bureau of Ships will tax our staff and facilities for some time to come," reported Rouse to Dean Dawson. "It is becoming essential that plans be made for expansion of laboratory and equipment." And indeed, in 1948, what remained of Nagler's turbine display was hauled away from the green across Riverside Drive and replaced by the institute's first research annex, the Model (later renamed West) Annex. That same year, the Bureau of Ships contracts were transferred to the Office of Naval Research (ONR), formed by an act of Congress just after the war to support diverse types of basic research and continue the federal government's wartime alliances with the scientific community. ONR immediately sponsored the construction of a new variable-pressure water tunnel, a low-velocity air tunnel, and even a scour flume and a sediment transport flume. The Navy's financial commitment to the institute expanded, and in agreement with the agency's mandate, ONR stepped in with a broad program to fund aspects of the institute's fundamental-research program such as its ongoing sediment studies. Rouse had reported to Dean Dawson that regardless of the low wartime publication rate, the experiences of World War II would be "of considerable value in our post-war research program." Little could he have predicted how those words would be played out financially as well, boosting IIHR's fundamental-research program in a manner that would serve Rouse well throughout his coming decades as director.

10. Historical Studies

IIHR's research and teaching activities are performed behind walls of brick and steel that occlude any vision of the importance of those activities to the general public. Yet the goings-on in the Hydraulics Laboratory are crucial to the understanding and management of water, which is essential for the survival of humans, societies, and the earth's functions. Many who have worked here have considered the impact of their efforts within the broader context of society's trends and needs. Their expanded perceptions are perhaps best exemplified by the institute's historical studies, which, although unusual for an engineering or scientific laboratory, have been integrated into the institute's purview and course work and have always constituted an important focus for its research and publications.

When Hunter Rouse arrived here in 1939, he brought with him a love of history and an appreciation of the written word that he had developed a few years earlier as a young professor at Columbia University. A decade later those interests coalesced around one of Rouse's graduate students, Simon Ince, who launched Rouse's love of history into a concerted research effort. Ince, who was Turkish, claimed that the immaculately barbered Rouse thought of him as "different" because he wore a beard and because Ince had composed a perceptive critique of functions and values in the American university that had been published in *The Transit*. Rouse thus approached Ince, whom he later described as "a student with considerable linguistic ability and interest in man's past and future," and convinced him to prepare a history of hydraulics for a doctoral dissertation. The result was an institute report entitled *A History of*

Hydraulics to the End of the Eighteenth Century. Rouse then spent a year as a Fulbright scholar at the University of Grenoble, where he combed through a collection of original hydraulics texts and created an expanded, reorganized, and rewritten book titled *History of Hydraulics.* The French journal *La Houille Blanche* published the chapters bilingually, as installments, in 1955. The complete book was published first in 1957, then reprinted in 1963 and 1980 and translated into Spanish and Japanese. It is being translated into Italian and remains a standard reference in the field.

As so often happens, success in one area leads to attempts in related areas. With the income from sales of *History of Hydraulics,* supplemented by moneys from other publications and from the University of Iowa, Rouse set out to create a comprehensive collection of historic hydraulics publications. Assisted by Frank Hanlin, a bibliographer at the university's Main Library, Rouse spent over a decade locating and purchasing books written by authors from Aristotle to von Karman, and Archimedes to Rouse himself. He obtained original works and first editions whenever possible. The result is a collection of over 500 rare or great books that inspires awe even in the novice. Among other things, Rouse's History of Hydraulics collection includes a gilt-edged volume of Pascal's 1663 treatise on the equilibrium of liquids; efforts by Galileo, Robert Hooke, and Leonhard Euler; and several books from the 1700s written by one or another of the Bernoullis, as well as works by later authors—Stokes, Reynolds, Prandtl, and others who were especially instrumental in elucidating underlying principles and outlining modern-day approaches to fluid motion. The books contain fine-lined drawings of river channels, flumes and gauges, water in various vessels and configurations, and water-guiding devices bedecked by chubby classical cherubs and mythical figures. The vellum covers, the yellowing pages, the cracking bindings, the very magnitude of the collection induces a sense of the weight of mental effort as well as artistry that have been invested in developing our current theories and constructs about water.

This collection was secured in the university library's Special Collections shelves because of its financial value and irreplaceability. However, Rouse had collected these books to use them. In 1960, Rouse inaugurated a course in the history of hydraulics, using his book collection as the students' primary reading and research mate-

Hunter Rouse's comprehensive collection of hydraulics texts, with many ancient and rare books, has been called the most significant book collection on the history of hydraulics in existence.

rials. Rouse's doctoral students were required to take the class, and through it to prepare original monographs either on the contributions of three (later two) major hydraulicians or on the historical development of a certain body of hydraulic knowledge. Students had to read relevant research in the author's original language. The catch here was that the original language of at least one hydraulician had to differ from the original language of the student. Thus, doctoral students in the Department of Mechanics and Hydraulics were given an intense, forced dose of history and foreign language simultaneously, and Rouse collected an ever-growing stack of papers on significant hydraulicians and historic developments.

A few of these students went on to publish historical papers of their own. Two translated books by Johann and Daniel Bernoulli. Many others assumed that Rouse would at some point use the student papers to expand the *History of Hydraulics* text. Perhaps such was the intent, but Rouse never did so. However, Rouse did continue to find pleasure in historical topics, and his historical research and writing expanded to fill his later years. "After one has reached what the French call a certain age, one becomes increasingly inclined to reminisce and eventually to look backward far more than forward," he explained in his 1972 article, "Hydraulicians are Human Too!," a series of sketches of significant twentieth-century hydraulicians whom Rouse had known. In 1976, the year of his retirement, the institute published Rouse's second text on the history of hydraulics, *Hydraulics in the United States 1776–1976*. This book, like its predecessor, emphasized the people behind critical events. Reviewers criticized the book for its lack of interpretation and analysis and for its "narrow definition of hydraulics" but found its great detail and completeness noteworthy.

Rouse then settled down to preparing shorter works on the subject. Some of these were presented as lectures at technical meetings that he continued to attend and were later published. A historical sketch of "how hydraulics . . . and related fields achieved the position they now hold at Colorado State University" appeared in 1980 in a booklet that Rouse edited on that subject. Other such talks and articles focused on significant publications, noteworthy hydraulicians, the evolution of hydraulics and fluid-mechanics research in the twentieth century, and professional organizations. Many of these, along with Rouse's other types of writings, have subsequently been compiled in the second volume of *Selected Writings of Hunter Rouse II,* which was published by the institute in 1991.

At age 77, several years after retiring to Sun City, Arizona, Rouse returned to the institute for a month to prepare an annotated listing of his history of hydraulics collection. This was published in 1984 in his last book, *Historic Writings on Hydraulics*. Rouse then left Iowa City for good, but his history books and book collection remain, worthy tributes to one of the few who has researched the history of hydraulics and a fitting legacy for his followers.

Although Rouse's love of history was exemplary, he was not the first at the institute to exercise this interest. Floyd Nagler had also sensed a passion for events of past years. Nagler's interest in history was nurtured quite naturally by his love of rambling through the countryside. While tracing streams hither and yon during his river surveys, he would occasionally come across millstones, wooden water wheels, or other such remains of earlier years, when water-powered mills had cut the rough lumber and ground the grain that made Iowa's earliest settlement possible. Nagler developed a fascination for these objects and their successors, the metal turbines driven first by water, later by steam.

With an energetic man like Nagler, fascination easily led to action. He was a master at mixing pleasure with work. When wandering about the state on consulting jobs, he began to collect the remains of the water mills. He returned to the Hydraulics Laboratory with millstones and turbines that he set up in front of the building, facing the street for all to see. When the Hydraulics Laboratory was expanded to its current size in 1932, Nagler insisted that a pair of millstones be inserted into the entranceway, "where the children can see them," as he explained, and where ever since they have continued to greet all who enter the laboratory. These particular millstones had been discovered by Nagler about ten miles north of Iowa City at the site of the pre–Civil War Hendricks grist mill, where they were facedown in the mud—a position credited with preserving the detail of their engraved faces. Nagler's positive evaluation of possibilities of damming the adjacent Mill Creek contributed to the future creation of Lake Macbride and Lake Macbride State Park at that site.

Nagler was not content with these expressions of his passion. His consulting work for the U.S. Army Corps of Engineers had provided him with two field teams that were supposed to help Nagler assess the rivers of eastern Iowa. Nagler asked these teams also to locate and photograph early mill sites. If the mill had succumbed to time and decomposition, then whatever remained would do. In this way, and during his own wanderings, Nagler collected a memorable (and now irreplaceable) set of photographs and slides of Iowa's early mills and their remains, which he used to illustrate a lecture on "The Passing of the Old Mills in Iowa." This talk was reportedly very popular and was repeated to a variety of civic groups in eastern Iowa. Nagler's enthusiasm is credited with inspiring Jacob Swisher's 1940 publication *Iowa, Land of Many Mills*, a book that remains the definitive volume on this aspect of Iowa's past.

Nagler's immediate successors at the institute, Mavis and Lane, did not adopt his passion for history, but neither did they totally abandon the subject. In 1936, Nagler's collection of water wheels was moved across the street to today's green space at the base of the limestone cliffs along Riverside Drive. A large explanatory sign

IIHR Collection, University Archives

Between 1936 and the mid-1940s, Nagler's collection of 13 turbines was displayed in "Water Wheel Park," the open space on Riverside Drive across from the Hydraulics Laboratory. Depression-era employees of Roosevelt's Works Projects Administration painted the moveable parts of the turbines silver and the stationary parts black.

titled the educational display "Development of the Turbine Water Wheel." None of the historic turbines have survived. Some may have been collected as scrap metal during World War II. Whatever remained was either hauled away when a laboratory annex was built on the site in 1948 or carted behind the annex and left to rust into oblivion.

During this same time period, the institute began to perceive its own efforts as historically significant, and it started producing a stream of bulletins summarizing institute efforts. At first, these consisted either of simple descriptions of the laboratory or of compilations of research and thesis abstracts (see University of Iowa Studies in Engineering Bulletins 19 and 26). But by the time the institute was celebrating its twenty-fifth anniversary, a more synthetic summary of the institute's history, facilities and equipment, classwork, and research in each of a dozen subject areas was published as a small booklet (Bulletin #30, published in 1946). It was followed by similar bulletins published in 1949 (#33) and 1960 (#40), and by Bulletin #44 in 1971 which constituted the nearly-200-page book, *The First Half Century of Hydraulic Research at the University of Iowa*. The current book, which thus follows a trail of similar publications, is the successor to that much-used 1971 book.

In the 1960s, Enzo Macagno extended IIHR's historical efforts into studies of Leonardo da Vinci. Macagno was becoming increasingly intrigued with the writings, drawings, and significance of da Vinci. Much has been written about this man's genius as an artist, engineer, and scientist. However, Macagno has advanced the thesis that da Vinci's most original work was in the field of flow and transport phenomena. Leonardo's efforts in this field were recorded in his codices and notebooks, where he wrote and sketched a disorganized and sometimes chaotic series of thoughts, theories, and experiments regarding a multitude of subjects. Leonardo's codices and notebooks—peculiarly handwritten backwards and in da Vinci's own cryptic version of the Tuscan language—remained unpublished and in fact were lost for several centuries after his death. But Macagno, who has trained himself to read fluently this special "Leonardian Italian," has been able to read the original writings of Leonardo, as well as the works of other scholars in their original language.

Combining his knowledge of language with his expertise in fluid mechanics, training in the humanities, and devotion to historical pursuits and interest in art, Macagno has been able to coordinate several themes dealing with fluid flow that are expressed in scattered fashion throughout the codices and notebooks. These he has developed through performing some of the experiments expressed in Leonardo's sketches—which require knowledgeable reading, interpretation, and coordination with surrounding text. Macagno also has integrated classroom work, explaining that his students "helped me unknowingly by answering quizzes tailored to discover primitive notions or reactions to questions that were considered by Leonardo." Macagno, once in conjunction with his wife and colleague Matilde Macagno, has recorded his translations of the codices and notebooks, his analysis and interpretation of their flow concepts, and his historical investigations in kinematics in a series of 17 monographs published by the institute between 1988 and 1998. His historical articles date from the early 1970s into the 1990s. He also has served as coeditor of the series Classical Works in Fluid Mechanics and Hydraulics, in which landmark papers have been reprinted in the journal *La Houille Blanche.*

In 1997, more than 40 years following his first University of Iowa appointment in 1956 and more than a dozen years after his retirement, Macagno continues to work on these topics. Matilde Macagno

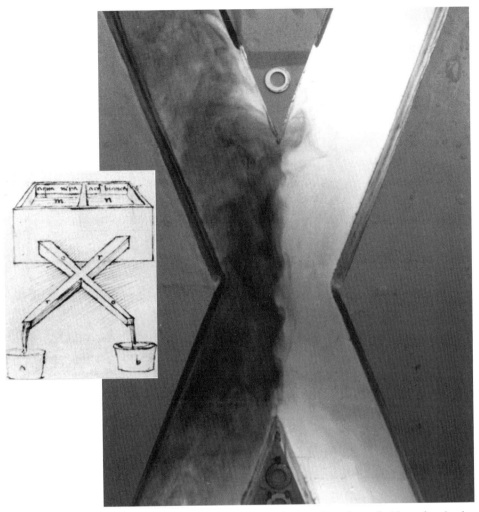

Enzo Macagno has investigated Leonardo da Vinci's thoughts about fluid mechanics in part by repeating some of the experiments sketched in his codices. Here, for example, Leonardo's sketch (left) of an experiment dealing with the amount of blending of fluids brought together by two conduits has been repeated in a modern experiment (right) with water of two colors brought together in open channels.

sometimes collaborates with him. An emerita professor of the university's Mathematics department, Matilde Macagno has for many years been engrossed in studies of the movement of water and has assisted with various institute research efforts. A primary interest has been her studies of the depiction of water and water movement through the ages by artists, engineers, and scientists.

With both Nagler and Rouse having set the stage for institute directors pursuing historical studies, it wasn't surprising that Rouse's successor John F. Kennedy commenced his own historical investigations in the mid-1980s. Kennedy was on sabbatical at the Swiss Federal Institute of Technology in Zurich, Switzerland, when he first wrote letters to the remaining family members of the century's greatest hydraulicians, beseeching them to share their reminiscences of the men who had shaped the studies of moving fluids. Kennedy's goal was a book explaining the modern history of the subject through a sequence of personal biographies, each touched with the intimate flavor and personal details of the hydraulician it described. This project, along with others, unfortunately yielded to the cancer that cut Kennedy's life and efforts short. But other efforts sprouted along the way. Kennedy did produce lengthy biographical sketches of Hunter Rouse and three other notable twentieth-century hydraulicians, Garbis Keulegan, Giulio De Marchi, and Albert Shields. Each of these is flavored with personalities, incidents, and anecdotes, and presents a fact-laden and fascinating story, just as Kennedy had hoped his entire book would do. And his letters to the families of hydraulicians gave birth to another institute project that continues to this day: the Hans Albert Einstein project.

Kennedy's letter to Einstein's widow, Elizabeth Roboz Einstein, reaped a rapid and warm response: not only did the octogenarian Elizabeth have many memories of her deceased husband, she had an entire book manuscript about him that she herself had written. Would Kennedy like to assist her in getting it published? While such an invitation might beg dismissal, Kennedy forged ahead and responded positively to Elizabeth. He had, after all, met Einstein and been a fellow student of the apparent whims of sediment transport. Also, what inquisitive person could refuse: Hans Albert was the son of Albert Einstein, the century's scientific figurehead and a man who, decades after his death, continues to inspire awe and curiosity. Elizabeth's manuscript made full use of this family connection but confused events and personalities. Despite the mixed quality and accuracy of Elizabeth's manuscript, Kennedy remained devoted to aiding her publication efforts. In 1990, he hired Connie Mutel to correct and complete the editing of that manuscript, which was published by IIHR in 1991 as Elizabeth's memoir, *Hans Albert Einstein: Reminiscences of His Life and Our Life Together*.

The institute then commenced a second biography of Hans Albert Einstein. The production of that book was stunted by Kennedy's death in 1991, but Robert Ettema, another institute lover of historical studies and writings, volunteered to carry on with Kennedy's scientific contributions to the text. Ettema and Mutel are continuing their efforts on this scientific biography, which promises to meld Einstein's life as the son of Albert Einstein with an assessment of his contributions to the study of sediment transport. With the Einstein project and Macagno's continuing efforts, as well as publications such as this very book describing the history of the institute and its hydraulic research, IIHR asserts its unusual eagerness to remain on the forefront of hydraulic research investigations, all the while simultaneously examining the history of hydraulic engineering and its progenitors.

11. Teaching and Outreach Activities

If Hunter Rouse was to play out his single-minded passion for infusing fluid mechanics into the thoughts and efforts of hydraulicians, students and practicing engineers would need to be educated about the fundamental principles of fluid mechanics. Rouse dedicated himself to this educational mandate with skill and ardor. "Hunter Rouse has been above all a teacher," his colleagues wrote on his 65th birthday, "one who has excelled in all of the many avenues a teacher can use."

Of course, Rouse did not initiate education as a major IIHR goal. Education has always been at the core of the institute's existence. Although a research institute, IIHR from the start has been part of the university's College of Engineering (then the College of Applied Science), the purpose of which is to educate future engineers. W. G. Raymond, first dean of the College of Applied Science, helped found the American Society for Engineering Education (then the Society for the Promotion of Engineering Education). Faculty members have always split their time between teaching responsibilities and the institute's research programs. Nagler and all succeeding professors held joint appointments at the institute and in one of the Engineering College's departments (at first, the Department of Mechanics and Hydraulics; then the Division of Energy Engineering; subsequently either Mechanical or Civil and Environmental Engineering). Through the years, the head of Mechanics and Hydraulics maintained close ties with the institute, sometimes (as with Professors Woodward and Mavis) moving back and forth between department head and institute leadership, fostering in both positions an environment in which teaching was enriched by research opportunities and research benefited from book learning.

Neither did Rouse initiate innovative teaching activities at the institute. The diverse material covered by institute faculty had always taken interesting forms. The institute's initial offering was an informal class in construction techniques: the first Hydraulics Laboratory was built largely by engineering students. A few years later, Nagler was instructing students in his Water Power Engineering class to design and build an educational model to demonstrate sources and applications of waterpower for display at the Iowa State Fair. Various other methods were also sought to convey information to the public. In the late 1930s, for example, the laboratory installed a permanent plumbing setup to demonstrate the dangers of water systems becoming polluted by self-induced vacuums. A few years later, demonstration lectures on plumbing dangers and sanitation were extended to medical students. And in 1939, IIHR held the first Hydraulics Conference, through it providing one of the only opportunities at the time for professional hydraulicians to meet and educate one another about their research.

Nagler, known for his fatherly concern for students, reached out to them in many ways, guiding their morale and character as well as their intellectual growth. Regarding the latter, he was the prime mover in establishing the graduate program in hydraulics. After persuading the Graduate College to grant research assistantships in hydraulics, he developed the necessary advanced course work. In 1922, under Nagler's tutelage, the first hydraulics students received their M.S. degrees. The graduate program in Mechanics and Hydraulics shot up tremendously under Mavis's tutelage in the 1930s, growing from eight students in 1935 to 26 two years later. In 1938, Iowa received the highest ranking of any American university for graduate education in hydraulic engineering during the previous five years, having produced twice as many graduate theses as any other U.S. institution. Already at that time the institute boasted of its ability to attract students from around the globe, with trainees from China, Canada, Mexico, Argentina, Germany, Hungary, Turkey, and India all studying here.

To educate those outside the institute about what was transpiring within, in the mid-1920s IIHR inaugurated an internally pub-

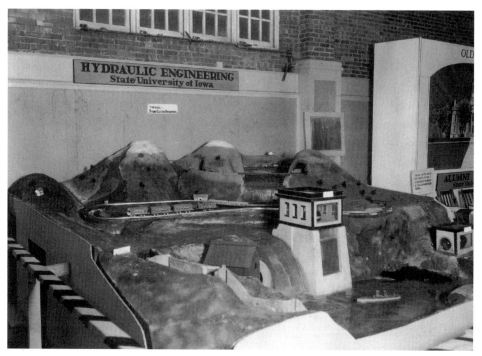

The Hydraulics Laboratory's innovative educational efforts began with projects such as this 1929 model, fabricated by students for a waterpower demonstration at the Iowa State Fair. The model included old and modern water wheels, locks, rivers, lakes, dams, an electric train, and transmission lines.

lished Studies in Engineering bulletin series. This too included the unconventional. In 1938, for example, one of the more charming bulletins was released, "The Road Map of Hydraulic Engineering in Iowa." This annotated map provided a touring guide for Iowa's rapidly growing contingent of automobile travelers—not to nature or cities or cultural amenities, but to nearly a hundred of Iowa's waterpower suppliers, channel improvement locations, pumping stations, and other hydraulic features. "The traveler interested in hydraulic engineering frequently passes within a short distance of a structure or project which he would wish to inspect if he were aware of its nearness. For him the Road Map of Hydraulic Engineering . . . has been prepared."

With precursors such as these, Rouse's focus on education cannot be called unique. He stepped into an institution with a tradition of providing innovative, high-quality education to a diversity of students. He joined others who were dedicated to raising a new crop of creative hydraulicians. Yet Rouse's focus on education was as vigorous and intense as anyone's could have been. He explored diverse teaching technologies as much as any other institute professor, if not more. In addition, he took the teaching of hydraulics to new levels by infusing it with principles of theoretical fluid mechanics.

Rouse took the reins of the institute as World War II was reducing Europe's laboratories and universities to a shambles. Students, turning their gaze to this country for leadership, support, and facilities, began to find what they needed at Iowa. Their numbers expanded even more as Rouse began to produce texts and films that became well known abroad. As the renown of Rouse and the institute spread, foreign students flocked here to study under him and then returned to their homelands, taking his ideas and approaches abroad through their own activities. Through them, and through his writings, lectures, films, newly initiated courses, interactions with students, designs for teaching laboratories and equipment, conferences, and administrative efforts, Rouse affected students and the teaching of hydraulics around the globe. For these many reasons, an exploration of Hunter Rouse as teacher serves as a prime example of the institute's educational efforts.

In the 1930s, when Rouse first came to IIHR, the leading technical schools were just starting to replace elementary hydraulics courses with instruction in general fluid mechanics. Until then, education in hydraulics had taught students the practical techniques that they would be applying as professional engineers. In 1938 (the year before Rouse's arrival), Iowa's Mechanics and Hydraulics department offered a roster typical of the times, with classes in hydraulics, hydrology, and hydraulic research and design. These taught students to describe and measure water flowing through orifices, pipes, and open channels and over weirs. Other classes offered training in irrigation techniques, turbines, river and flood control, etc. A single graduate-level class, Technical Fluid Mechanics, educated students about the physical properties of fluids.

This applications-oriented approach was anathema to Rouse, who (unlike many others at the time) insisted that his students understand the broader theories underlying their experimental work

Rouse insisted that IIHR's teaching facilities (such as this cavitation tunnel) be "designed to illustrate principles of fluid mechanics rather than merely to establish calibrations of equipment." The tunnel was developed to show the vaporization of water when its pressure drops to vapor pressure as flow accelerates through the constriction.

and professional activities. "The Institute's policy . . . toward teaching has been, and continues to be, directed toward the study of fundamental principles," stated a summary of the institute in 1960, after Rouse had kneaded its direction for 15 years. "Research and instructional activities are designed to complement each other." Questions of a fundamental nature were the keys for unlocking the functions of the natural world; they were the gates to understanding the movement and response of fluids so intimately that one could understand and predict future events. Fluid-mechanics theories were to pervade all aspects of the university's education in hydraulics: the classroom, the teaching laboratory, the choice of research topics by both students and professors. Everything was to work together to promote the discovery and dissemination of fundamental knowledge. The reliance on traditional empiricism, rote testing, and product development all must go.

Upon arrival at IIHR in 1939, Rouse immediately initiated a general reorganization of classes, instructional methods, and equipment. During his first year, he transformed the Hydraulics class into Mechanics of Fluids, and the Hydraulics Laboratory into the Fluid Mechanics Laboratory. Graduate students could take Rouse's Advanced Fluid Mechanics, which replaced Mavis's Technical Fluid Mechanics.

With U.S. entry into World War II, it became evident that the number of U.S. engineering graduates was not sufficient to meet the needs of national defense. Iowa's Engineering College, along with other schools across the country, was pulled into the federally subsidized National Defense Training Program, through which engineers were prepared to fill specific technical positions in defense industries. Here again Rouse made his mark. While most short courses dealt with the practical applications such as the chemistry of explosives, radio engineering and construction, or airport engineering, Hunter Rouse taught his already intensive fluid-mechanics class in an even more intensive, four-week format. Students fully earned their four graduate credits by filling their days and evenings with Rouse's demanding lectures, laboratory sessions, problems, and readings.

The war brought out the shortcomings of engineering education generally. Although engineers had made contributions to the development of ships, tanks, planes, and armament, physicists with advanced fundamental training had taken the initiative in creating new devices and systems. Thus, the need to train graduate engineers in the physics of fluid flows and other fundamentals became obvious, even while wartime federal funding was expanding research and training opportunities for graduate students. These changes occurred in tandem with Rouse's continued revision of class offerings. He and Lane had been teaching Advanced Hydraulics Laboratory. In 1946, the year Lane left Iowa, Rouse renamed the class Advanced Laboratory Investigation and substituted demonstrations of fluid-mechanical processes for exercises in model work and theories of hydraulic similitude. He introduced an Intermediate Mechanics of Fluids course and, in 1950, an Intermediate Fluid Mechanics Laboratory. The institute now had a three-tiered fluid-mechanics series, with classes and labs at the beginning, intermediate, and advanced levels. By 1960, both Intermediate Fluid Mechanics and Advanced Fluid Mechanics had expanded into a two-semester sequence, and a seminar, Topics of Mechanics of Fluids, had been added. Rouse both initiated and taught (sometimes with Emmett Laursen or John McNown) the intermediate and advanced level classes and labs. In 1960, Rouse also started teaching History of Hydraulics "in the belief that students attaining the doctoral level should know something about the background of their profession."

Rouse's ideas about education were quite firmly established. He was interested in students and institutions striving to reach the top, whose dream it was to take hydraulics into worlds yet unknown. That, in turn, implied certain things about class work, research, and collegial communications. "The art of advancing any field of knowledge necessarily involves fundamental research, high-level education, and free intercourse among those who are engaged in such endeavors," he wrote in a paper on advancing the science of hydraulics. Research, class work, and communications were all to be directed toward the education of the few leaders, not the more task-oriented training of the majority of students. This idea was implanted in him as a young man, when he had studied in Germany and been impressed by the quality of the country's technological institutes. Those schools accepted only the most talented young men and groomed them into a new crop of leaders. The training of the more numerous students, who would execute routine engineering details or technical tasks, was left to other institutions. While Rouse's writings routinely admonished the U.S. university system for "unfortunately . . . attempting to train the masses as well as educate the leaders, simply because every student wants a college degree regardless of ability or ultimate function," he refused to lower his standards for his own students and institute.

Instead, he continually imposed his high expectations on the institute's student body. Leave it to others to apply knowledge; Rouse was training students as planners and thinkers who developed knowledge. His lectures were models of organization, fast moving and content-packed. Rouse, in preparation, would write them out and then rip up his notes, delivering each meticulous lecture from memory. The flow of thoughts and equations would be neatly inscribed on the blackboard, which students would scurry to copy after class. The lectures assumed a level of knowledge that, if not possessed, would leave students scrambling to catch up. Rouse neither elaborated nor wasted words. Students learned not to ask for repeated explanations if they did not understand a point; Rouse never explained twice and always moved forward. The rapid pace induced students sometimes to sit through Rouse's classes for a few additional semesters after they had taken them for credit.

The class sessions could be trials for even the best prepared. Rouse presented the information and then expected the students, through rigorous analysis, to come up with all interpretations themselves. In the process, they suffered through Rouse's constant questioning. He carried with him a deck of note cards with his students' names on them. Shuffling the cards, he would ask a question and start at the top of the deck, working through the deck and asking the same question without comment until he got a correct answer, then another question that, like the others, was constantly pushing students to expand and defend their thoughts and to clarify their ideas. Rouse's multiple-choice tests required a similar depth of understanding. They had to be read with extreme care to detect the minute point that would separate the one correct answer from those that were incorrect. Students would collect old examinations and use them to practice this skill.

To survive a Rouse class demanded total commitment and also a tough skin. When someone was not meeting expectations, Rouse was not kind. One graduate student, breaking down after his oral exam, told of Rouse announcing that the student's presumably complete answer to a certain question was "the worst answer I've ever heard." Another complained that "whatever I was supposed to be mastering, I wasn't mastering it well enough to avoid Hunter Rouse's sharp tongue." Students were fearful of this exacting, demanding teacher, to the point that some could not speak a word to his face. Others were humiliated or angered by his remonstrations. The class demands also were grueling. Students would sacrifice family for books and problems late into each night. Those involved with research were expected to be in the laboratory every evening and weekend. Mistakenly expecting a break over the Christmas holidays, they would find themselves instead tediously drawing large flow nets by hand, the fishnet-like mesh of intersecting lines tracing the flow of water over a weir or around an obstacle.

Yet students would arrange their course work so they could take Rouse's arduous classes, perhaps sensing that in later years many

would consider these the pinnacle of their education. As one graduate student later wrote to Rouse, "I worked harder on your course in
Fluid Mechanics than on any other, but I got the lowest grade. . . . In
retrospect what I learned . . . has had a greater influence on my subsequent career than any of the other courses." Students found that
by not being allowed to cut corners, they came to a deeper understanding of fluid mechanics than they thought possible. Rouse was
fair, and if one followed his rules, one survived. He also let students
know exactly where they stood with him. In later years, they found
that the mark of his expectations carried over. "He held us to higher
standards, I'm sure, than any of us thought we could ever reach,"
quoted one Rouse-student-become-dean. Students developed a tremendous respect for this mentor who, many sensed, was setting the
course of their career. "Even though we had to pass through some
awkward moments trying to decipher the solutions to the problems
you posed in class, slowly we realized that, in the process, the subject was getting etched in our minds," wrote another student to
Rouse many years after graduation. "I have never come across a
teacher like you who, by sheer intellect, depth of knowledge and the
capacity to provoke deep thinking in students, has become legend
in his own lifetime."

When Rouse started teaching fluid mechanics, he had a wealth
of German texts to choose from, but almost none in English. Rouse
thus had to prepare his own. He had learned to take copious class
notes during his classwork in Germany, where no textbooks were
used. Thus, it was nothing new for him to prepare and duplicate a
set of notes for students to use during his first year of employment
at MIT, where he was assisting Professor Spannhake to teach an applied hydrodynamics class. These notes, combined with materials
he prepared for his own teaching, first at Columbia University and
then at California Institute of Technology, fed directly into his first
book, *Fluid Mechanics for Hydraulic Engineers* (first published as an
Engineering Societies Monograph in 1938, republished by Dover in
1961). In this innovative text, Rouse combined experimental hydraulics with understandings of fluid-mechanics principles whose
development had been stimulated by the rise of aviation and applied aeronautics. In the book's preface, he explains that while there
were numerous elementary fluid-mechanics texts for hydraulic en-

gineers, the only texts for more advanced courses used aeronautical applications. Rouse's text was the first advanced fluid-mechanics text specifically for hydraulic engineers—the first to "place at the hydraulician's disposal a group of principles and concepts that had already proved their worth in related fields." In it, "every phase of the subject is developed from basic principles of mechanics," rather than the empirical methods that had been the mainstay of hydraulic engineering for centuries. Rouse hoped that his book would "convince the hydraulic engineer that there is much to be gained by giving heed to rational methods of analysis and research."

Rouse had found a book-writing pattern that worked. During his second year at Iowa, in 1940, he developed notes for an undergraduate fluid-mechanics course and started to test them on students. These notes fed into his second text, *Elementary Mechanics of Fluids*, published in 1946 by John Wiley and Sons (and subsequently translated into several foreign languages and republished by Dover in 1978). This book was one of several similar fluid-mechanics texts published in the 1940s. Undaunted by the fact that it appeared in a single edition (while its competitors lived through several), Rouse went on to produce two more texts: *Basic Mechanics of Fluids* (co-authored with Joe Howe and published by Wiley and Sons in 1953), and *Advanced Mechanics of Fluids* (edited by Rouse, published by Wiley and Sons in 1959). *Basic Mechanics* was similar to Rouse's earlier *Elementary Mechanics* text but intended for beginning undergraduates. *Basic Mechanics* was very widely used as a text in this country and remains a valued professional reference to this day. *Advanced Mechanics*, a sequel to the *Elementary Mechanics* text, was a group writing effort of several IIHR professors who took a decade to produce the manuscript. It was translated and used as a text abroad as well as in this country. In fact, Rouse's texts (as well as his journal publications) became recruitment tools, familiarizing students in other countries with Rouse and IIHR and drawing them to study here from the "master."

Through Rouse, IIHR became known as an important source of hydraulics texts. During these middle years, IIHR also became a source of other books on hydraulics, including Baines's *A Literature Survey of Boundary-Layer Development on Smooth and Rough Surfaces at Zero Pressure Gradient* (1951) and *Wind-Tunnel Studies of Pressure Distribution on Elementary Building Forms* (Chien et al., 1952), which were published by IIHR. Lou Landweber edited the translations from Russian of *Theory of Ship Waves and Wave Resistance*, published locally in 1968, and the two-volume set *Theory of Ship Motions*, published by Dover in 1962. Two students involved in Rouse's History of Hydraulics class translated from Latin the eighteenth-century treatises *Hydrodynamics* and *Hydraulics*, by (respectively) Daniel and Johann Bernoulli, which were published by Dover in 1968.

When Rouse started to teach fluid mechanics at Iowa, he found that the traditional hydraulics teaching laboratory experiments and equipment needed to be revamped. He started to institute the necessary changes. "Hydraulics was long taught in American engineering colleges as a vocational rather than a scientific course," Rouse wrote in retrospect. "Experiments involved the use of gages, the calibration of flow meters, the measurement of head loss in a pipe, and performance tests on a pump or turbine." While there was nothing wrong with teaching such matters, "the error lies in disregarding the fact that [the engineer] needs to know a great deal more. . . . The true university . . . curriculum emphasizes grasp of scientific principle rather than practice in manipulating equipment."

Rouse's efforts in laboratory design and construction had commenced while an instructor at Columbia. He went on to initiate the first formal fluid-mechanics teaching laboratories at Iowa. "Laboratory equipment can be designed to illustrate far more than manipulation," Rouse stated, and he proceeded to design, construct, and test appropriate equipment at IIHR. He designed, for example, the complex brass pipe system that still occupies the southwest wall of the Hydraulics Lab's first floor (see figure). This system, with its various lengths and diameters of pipes, its elbows, splits, and expansions, was constructed for students in the elementary fluid-mechanics course to learn about transition effects and energy distribution within closed conduits. IIHR's undergraduate and graduate laboratory facilities were steadily expanded. In 1948, much of IIHR's experimental equipment was moved into the new Model (later West) Annex, allowing the teaching laboratory to double in size and take over the entire central portion of the first floor of the Hydraulics Laboratory. A considerable amount of new equipment was added.

In the mid-1940s, Iowa's teaching facilities started to serve as models that were copied by other universities. By 1945, the National

Rouse instituted IIHR's first fluid-mechanics teaching laboratory and developed the equipment that it held, such as this pipe system.

University of Colombia at Bogota had contracted with IIHR to design a three-story hydraulics laboratory for both instruction and research, complete with equipment, and then to supervise its construction, installation, and initial operation. Rouse himself made the preliminary design sketches, with the result that the completed facilities would be "far more modern than those of the Institute itself." In 1948, a similar project was initiated by the Central University of Venezuela at Caracas. IIHR's experiences continued to fuel education elsewhere as equipment was shipped from IIHR to several U.S. universities and as others reproduced equipment that was based on IIHR's designs and specifications. In 1954, IIHR received a contract to design a laboratory for the University of the Philippines in Manila, and in 1960 Rouse traveled to universities in the United Arab Republic to advise them on modernization of their laboratory instruction. Finally, in 1961, in order to further "laboratory instruction on the same scientific level" as was occurring at Iowa, the institute published Rouse's 60-page manual, *Laboratory Instruction in the Mechanics of Fluids* (Bulletin 41). This outlined teaching instruments and equipment and their construction, along with instructional goals and procedures. The manual proceeds in practical fashion from the basics ("the most important requisite . . . is uncluttered space") through matters such as effectively sealing flumes, coupling pipe assemblies, and laying out instructional experiments in cavitation. It has been widely used for designing research and instructional facilities both in the U.S. and abroad.

Laboratories provide hands-on experience to teach what is part and parcel of all engineering experience. But what about schools with limited facilities and financial resources? For these, Rouse had another plan: provide them with movies of crucial experiments and processes. If the students couldn't perform the experiments or observe them firsthand, they could at least witness them on the big screen. With this goal in mind, Rouse started to prepare such films.

While Rouse developed more films than anyone else at the institute, his photographic efforts were not unique, especially if one considers the use of films in research. Back in 1929, Floyd Nagler had been making silent films on his studies of bridge piers and flow around bends. In the 1930s, the need for better visualization of flows had led to the use of clear water in model studies and transparent flumes and pipes. Taking another step toward improving visualization, the institute established its own darkroom, bought motion-picture cameras and editors and projectors, and started to produce films that used air bubbles, dyes, or aluminum flakes to allow visualization of motion in transparent fluids. A camera might be mounted firmly on a carriage above a flume, for example, and aluminum flakes scattered on the water. Under intense illumination, the camera would trace the light glinting off the flakes as they spun down the flume and swirled into eddies. Or comparing sequential photographs taken under stroboscopic light sources would allow researchers to measure the velocity of cavitation bubbles or sediment particles in turbulence eddies. Such photographic records of fluid phenomena were a common component of hydraulic research in those years. Films of certain phenomena became a routine part of

A speaker at the 1946 Hydraulics Conference claimed that "experimental hydraulics probably makes more use of a wider variety of photographic techniques than any other science." The use of movies for research and educational purposes began early at IIHR, as demonstrated by this 1934 filming of model tests of ship locks.

the research process and were regularly submitted to granting agencies as part of a project's final report.

Soon, films were drawn into educational use. During the war, the institute produced films on air turbulence and diffusion of gases in its new air tunnel that were used for training Chemical Warfare Service military personnel. The practice soon became routine. In 1945, IIHR was preparing motion pictures on venting problems of small houses. Rouse used specially prepared motion pictures to supplement laboratory demonstrations in his fluid-mechanics classes. And for many years, institute personnel were sufficiently advanced in the use of films to advise others on their execution. In 1937, for example, Chesley Posey published a paper on the production of 360-degree panoramic motion pictures in a motion-picture engineering journal.

With time, filming projects were turned over to Fred Kent, Iowa City's preeminent photographer and head of the university's photo service. That service assisted Rouse with subsequent films, the first being a documentary on IIHR started in 1951. Rouse used this 20-minute film explaining the institute's research and mission when he presented talks on the institute both here and in foreign countries. He also showed it on television. Through the years, he made other motion-picture sequences displaying fluid properties and integrated these regularly into talks given at professional meetings and to the general public. One of many examples was his 1942 talk to the Iowa City populace at large, "How Fluids Flow—A Study in Kodachrome," that used color motion pictures in which he had "pictorially captured the action of flowing water by means of dyes, air bubbles, and aluminum particles"—a feat in those days worthy of special regard.

In 1960, Rouse received NSF funding to produce an extensive educational film series. MIT likewise received funding for a similar project. Rouse wrote the scripts for IIHR films. His trusty shop manager, Dale Harris, supervised fabrication of "one after another of the previously unheard-of props demanded by the script." All six films were narrated by Rouse, who appears in his standard form, without a flaw in personal appearance or language. Although delivered like rapid-paced lectures, Rouse speaks without hesitation or reference to notes, formally, elegantly, and precisely, in perfectly constructed and executed sentences. As in life, Rouse presents an image of absolute control.

Rouse was particularly proud that his films used color throughout, a feature that distinguished them from the competing MIT's films which, Rouse wrote, were "rather slow-paced." Rouse's were far from that. If anything, the six 20-to-30-minute films move extremely rapidly through the phenomena treated in the usual undergraduate fluid-mechanics courses: introductory concepts, fundamental principles, gravitational effects, laminar and turbulent flow, lift and drag, and compressibility. The movies use a number and diversity of visual displays that are sophisticated even for the present day. Scenes of scuba divers and weightless, floating cosmonauts follow everyday scenes of persons washing their hands or stirring coffee. Flights of parachuters switch to shots of ocean waves, mountain fog, or balloons against a blue sky, and rapidly back again to ships at sea. Equations without derivation and animations magically appear on the screen; however, lectures that fully explained concepts and derivations were not the intent. Instead, these films were to supplement standard lectures and laboratory demonstrations that could never portray the variety of flow phenomena displayed on the screen. These films, which required nearly a decade to complete, were to provide the illustrative material that made classroom lectures meaningful. Through the films, any teaching institution, here or abroad, could show students a large variety of fluid phenomena, including those in natural settings, regardless of laboratory size or sophistication. Students around the globe could hear an international expert explain fluid flows. The number of films sold—by 1976, 425 copies of one film or another had gone out to 150 institutions in 35 countries—shows that the films achieved this goal. They were even shown to science classes in Iowa's high schools. In 1985, they were reissued in videotape format, and copies are occasionally ordered to this day.

Rouse's films offered education to distant students and a tool to their teachers, but in the 1940s, there were few opportunities for practicing hydraulicians to educate each other by sharing their research results and ideas in person. IIHR helped to fill this gap by organizing and hosting the Hydraulics Conferences, a set of seven professional meetings held at three-year intervals (excluding wartime delays). The first such conference was the 1939 brainchild of F. T. Mavis, former associate director of IIHR, who then chaired the Department of Mechanics and Hydraulics. Mavis's conference, sponsored jointly by the Society for the Promotion of Engineering Education, consisted of four days of invited presentations on a variety of hydraulics topics (models, sediment transport, turbulence, hydrology, etc.) that were later published by the institute.

This first conference, with its enthusiastic audience of 250 participants from widely scattered locations, was deemed a success. The organizing and hosting of its successors proved to be one of the institute's major efforts during these years. All conferences consisted of invited presentations, later published as proceedings, but Rouse made one major alteration: he suggested a single theme for each conference to encourage in-depth discussions. These themes covered applications of fluid mechanics in many interrelated fields (1942; Bulletin 27), post-war applications of war research (1946; Bulletin 31), sediment transportation (1952; Bulletin 34), measurement of fluid flow (1955; Bulletin 36), and agreement between prototype and model behavior (1958; Bulletin 39). The 1955 conference was supplemented by a two-and-a-half-week course on flow-measurement techniques for those who chose to stay on.

The fourth Hydraulics Conference (held in 1949) broke this pattern. Instead, it was designed as a set of technical sessions for the creation of an authoritative and comprehensive reference volume on engineering hydraulics—a treatise that was noticeably lacking in this country. A tremendous amount of organization fed into the creation of the thousand-plus-page *Engineering Hydraulics*. Three years before the conference, the field was divided into 13 subjects or chapters—fundamentals of flow, flow measurement, wave motion, and the like—and authors were chosen for each chapter. Outlines were checked and chapters drafted, edited by the most exacting of editors (Hunter Rouse himself), and sent out to conference participants before they arrived in Iowa City. Each chapter was then critically reviewed and discussed at conference sessions, the chapters were revised and again edited, and the book was published by John Wiley and Sons. The massive amount of effort that fed into the book's production was merited by the book's acceptance. Books sold rapidly and well and were widely used, with 6500 copies selling in a decade, 2800 of those going to foreign countries.

These Iowa conferences were deemed significant as much for their opportunities to gather and discuss ideas as for their formal presentations. Their proceedings became widely dispersed publications, those from all conferences combined numbering some 11,000 volumes. They also must have outranked the stock modern conference in civility and entertainment. Dean Dawson hosted all incoming guests at the opening meal, a buffet dinner in his home. There were trips to the Amanas, Iowa, much socializing, and throughout the meeting the women who came along with their husbands were cared for and entertained by the Iowa City wives. However, for all their amenities, by 1950 IIHR's hydraulics conferences were no longer unique. That year the Hydraulics Division of the ASCE started to host its specialty conferences. While attendance at the

Hydraulics Conferences had risen from 250 in 1939 to a high of 425 a decade later, including participants from many foreign countries as well as numerous states, participation dropped to 173 by 1958. By that time, the proceedings were no longer widely distributed, and at least ten American conference series dealt with fluid motion—indeed four were scheduled within a few weeks of each other that summer. The series was thus terminated. A single swan song to the Hydraulics Conferences was sung in 1966, when an IIHR Hydraulics Colloquium allowed former graduate students to mingle for three days of roundtable discussions and social events.

Rouse's educational excursions did not in any way end with his departure from the IIHR directorship. Throughout his six years as dean, he continued to teach and to supervise graduate theses. Taking funds allotted for renovating the Engineering Building, he moved the Engineering Library into larger, air-conditioned, newly remodeled quarters, transforming it into the college's showpiece. He also stimulated thoughts by hosting interdisciplinary symposia for engineering alumni on "Technology and the Spirit of Man" and printing the talks. His prolific pen started to produce writings on his educational theories, philosophies, and observations.

His most consuming educational effort, however, was the complete restructuring of the College of Engineering's undergraduate curriculum. The times were calling for a broadening of the engineering college's curriculum as a whole, and engineering schools across the country were responding by altering their required course work. Rouse wrote that the engineering student must now "be educated to take his proper place in modern civilization as an engineer-citizen" and take classes that would "fill his leisure hours with the best in music, art, literature, and philosophy," and also learn to "understand why and how other people think, speak, and coexist." This meant integrating the humanities into a more unified curriculum, balancing them with mathematics, science, and "the art of creative engineering design." The resulting major readjustments in required classes were guided by a faculty committee, in which Rouse's role was "stimulating progress and encouraging consensus rather than promoting my own ideas."

Rouse's educational excursions continued past his retirement from the University of Iowa. He had first taught an intensive fluid-mechanics course at Colorado State University during the summer of 1940, and in 1976 he started returning there each summer until 1987 to teach intermediate fluid mechanics and, later, the history of hydraulics. His high standards soon produced predictable results: Rouse volunteered to sit in on the graduate exams that were to be given during the summer, and soon no students scheduled their orals during that time. Rouse also continued his travels and teaching elsewhere, presenting lecture series in South America, Australia, Taiwan, and the People's Republic of China as well as in the United States, before his traveling urge started to quiet in 1985.

Nor did Rouse's departure deter IIHR's educational function. Classes continued to be developed and revised. Undergraduates and graduate students continued to move through the ranks, receive their degrees, and carry their IIHR training out into teaching, research, and practicing engineering positions around the world.

IIHR continues to disseminate information through the production of books, although these are no longer produced with the flurry that characterized the Rouse years. Between 1977 and 1980, IIHR professor Thomas Croley wrote three books, published by IIHR, on hydraulic and hydrological applications of programmable calculators and mini-computers. The books, which included programs and user instructions, became significant, widely used references. Other books include two volumes of Hunter Rouse's writings (*Selected Writings of Hunter Rouse*, Volumes I and II, published in 1971 and 1991), and in 1991 a memoir written by Elizabeth Einstein about her late husband the hydraulician Hans Albert Einstein, which was edited by Mutel and Kennedy. The institute has also reprinted and distributed classical books in hydraulics that are still of limited interest: *Estuary and Coastline Hydrodynamics* by Ippen; *Cavitation* by Knapp, Daily, and Hammitt; *Practical Aspects of Computational River Hydraulics* by Cunge, Holly, and Verwey; and *Three-Dimensional Turbulent Boundary Layers* by Nash and Patel.

With the termination of the Hydraulics Conferences, IIHR's energies along these lines shifted to organizing and hosting conferences for professional societies and editing proceedings of such meetings. Examples include the "International Symposium on Refined Flow Modeling and Turbulence Measurements" (1985) and the 1986 "IAHR [International Association for Hydraulic Research] Symposium on Ice." In 1992, IIHR helped organize the seminar

"Prediction and Damage Mitigation of Meteorologically Induced Natural Disasters," held in Taiwan. The year 1993 was especially busy: IIHR co-sponsored and hosted the "Sixth International Conference on Numerical Ship Hydrodynamics," held in honor of Emeritus Professor Lou Landweber, and edited its proceedings; chaired and hosted the "Fourth International Conference on Precipitation: Hydrological and Meteorological Aspects of Rainfall Measurement and Predictability," and edited its proceedings; and helped organize the "U.S.-Spain Workshop on Natural Hazards," held in Barcelona, and distributed its proceedings. In 1995, in honor of deceased IIHR director John Kennedy, IIHR organized "Issues and Directions in Hydraulics—An Iowa Hydraulics Colloquium." Invited speakers explored the major issues and future directions that characterize hydraulics at the end of the twentieth century. Its proceedings were published as a book in 1996 by A. A. Balkema of Rotterdam. In 1997, the IAHR Congress, "Water for a Changing Global Community," was organized through the institute.

The institute no longer publishes the Studies in Engineering bulletin series that Nagler initiated in 1926; the last such bulletin, #44, was published in 1971. However, the IIHR Technical Report Series, which first appeared in 1966, in 1997 was approaching its 300th issue. All of these reports have been written by IIHR researchers. In addition, a series of Limited Distribution Reports (initiated in 1971 and consisting of reports that are less widely distributed) by 1997 had surpassed #250. Another special series of 17 IIHR monographs focuses on analysis of the works of Leonardo da Vinci. Each of these monographs has been composed by Enzo or Matilde Macagno. This combination of IIHR publications, all of which are dedicated to information exchange among professionals, supplements the numerous papers and other miscellaneous publications written by IIHR staff that are published in professional journals, conference proceedings, and elsewhere.

Videotapes are still made at IIHR, but no longer for the classroom or general public as during Rouse's tenure. Today their use is limited to research. Frequently videotapes are included as part of a final contract or grant report, offering a visual interpretation of the performance of a hydraulic model, for example. The institute's video equipment is also hauled into use as part of the research process, contributing, for example, to the measurement of flow velocity.

In the 1990s, Rouse would easily recognize the institute laboratory that was being used for teaching. It was still located on the first floor of the Hydraulics Laboratory and included some of his original equipment. However, in 1997 a state-of-the-art teaching laboratory, possibly with a different location, was in the planning process. In addition to the traditional flumes and pipe assemblies needed to simulate fluid flow in nature, the new laboratory would contain computers and programs to simulate flows numerically, laser-based measuring instruments, and equipment for computer-aided data acquisition. Accessories (such as aquariums, films, pictures, gadgets, and other displays disbursed around the lab) as well as the choice of contemporary research topics would heighten student interest, make the subject of fluid mechanics visually attractive, and stress its relevance to everyday life processes. Techniques for assuring accuracy in both experimentation and data interpretation—topics often neglected in the 1990s—were being given priority. Although this laboratory would differ in certain ways from Rouse's, he was not to be forgotten. A historic display on his contributions to the field was planned. While this laboratory would be used mostly for introductory undergraduate classes, graduate students and courses would also use it on occasion.

A substantial revision of the undergraduate and graduate curricula in fluid mechanics, hydraulics, and water resources engineering was progressing simultaneously. To remain at the forefront of education, new courses such as hydroclimatology and remote sensing have steadily been added to the roster, and new techniques and instruments have been incorporated into class demonstrations. The faculty in 1997 was stepping back to assess the effectiveness of class offerings as a unit, both in attracting students to IIHR and in training them for the profession, and to adjust the curriculum as necessary. The newly designed graduate package was being publicized here and abroad through the World Wide Web.

Just as traditional methods of education continue to provide a major focus of institute activity, so too does innovation. Director V. C. Patel in 1997 instigated a novel program, International Perspectives in Water Resources Planning, designed to broaden the international dimension of an IIHR education. A large percentage of IIHR's graduate students—at least half—have always been from foreign countries. Among the 1997 faculty, over half were born

IIHR's first International Perspectives in Water Resources Planning trip held in January 1998 took students to India for 18 days with Professor Subhash Jain (seated, far right) to study the multiple aspects and impacts of the controversial Narmada Dam and related water-resources projects.

abroad, and most travel or consult abroad routinely. Such students and professors have always added a breadth of approach and experience to institute endeavors, but primarily through their words. Through the International Perspectives program, seniors and graduate students are able to learn about problems and solutions entertained in other countries directly, through their own experiences. For a period of two to three weeks, students travel to a given country to focus on a topic in water-resources planning. Site visits are combined with on-site seminars presented by experts in the historical, cultural, ethical, scientific, and other aspects of the given question. The first class, involving 11 students, was offered in January 1998 in India. Future classes will move to other countries and sites and may be opened to younger undergraduates and to students from other institutions. This on-site international training adds another new facet to IIHR's educational experience. Patel hopes that in addition to enriching the minds of participants, the program will draw in graduate students with broad perspectives and will redefine IIHR as a center for international training in hydraulics.

IIHR staff, 1975

III CUTTING NEW CHANNELS
The Last Three Decades 1966–1997

12. John F. Kennedy, Applied Hydraulics Studies, and Recent Changes

John F. Kennedy, who usually went by "Jack," had been director of IIHR only six months when, in January 1967, he wrote his first annual report concerning the institute's activities. He addressed it to his meticulous, exacting predecessor, Hunter Rouse, who had recently stepped into the dean's office of the College of Engineering. Kennedy's task must have been a daunting one. He was a scant 33 years old and just seven years out of graduate school. In contrast, Rouse, a man of stature and renown, was known as the "father of modern hydraulics" around the world. He was also known for his strict standards and sharp tongue. Many in Kennedy's position would have been tempted to follow the tried and true, writing that they had adopted Rouse's goals for the institute and followed Rouse's protocols.

Yet Kennedy clearly showed that he acknowledged few sacred cows. His report, although tactful, challenged Rouse's unilateral emphasis on basic research. "Both pure and applied research are essential, and it is equally important to strike a proper balance between the two," Kennedy wrote. While Rouse had stated that universities should limit their activities to pure research and leave testing and applications to others, Kennedy denigrated the "attitude abroad which seems to value research pursued for its own sake above that which has goals and applications." He compelled institute researchers always to look toward the "potential or realizable application of the results." Kennedy verbally denied the institute researcher the "luxury of pursuing his research whims without ever considering how the results bear on the multitude of immediate, important, unsolved problems faced by those who must make practical technological decisions." The following year, in his second annual report, Kennedy condemned even more strongly the emphasis on basic research that had dominated the preceding decades at IIHR and elsewhere across the country:

> When the technological history of the last twenty years is written it may well be characterized as the age of research affluence, when almost any type of basic research, and even unstructured, aimless investigation posing as basic research, could gain support. This era appears to be drawing to a close.

While never specifically accusing Rouse or his immediate staff of pursuing "aimless investigation," Kennedy clearly was calling for a change of approach. Thus, once again, for the third time, IIHR was being reshaped through a visionary director whose personal inclinations stepped in time with the needs of the day. In the 1920s, Nagler had established a firm research program with his energetic pursuit of governmentally and industrially funded applied projects. In the 1940s, Rouse had veered away from the applied, and infused IIHR's laboratories with government-funded basic research. Now Kennedy posed a third option, a synthesis of the two previous approaches: IIHR would pursue applied research initiatives, largely funded by industry, from which would spin a multitude of basic research initiatives that could be funded by government grants.

By the mid-1960s, when Kennedy arrived, IIHR was in danger of losing its world-status ranking, established largely by Rouse's prodigious efforts. While its contributions to basic research and education were prolific, its fiscal condition was becoming precarious. A small number of broad contracts from the Office of Naval Research provided the major support for all senior staff. Other rival laboratories, many organized and directed by IIHR alumni, were successfully competing with IIHR for research funds. Rather than growing and diversifying, IIHR's research staff remained small, with its research initiatives constant from year to year. This resulted in part from Rouse's proscription of model studies and site-specific investigations, which Kennedy observed was "one of the cardinal principles of Rouse's research-management policy." The result was perceived as a disheartening lack of vitality and innovation.

Further summarizing the stagnation that he sensed at IIHR

when he first arrived, Kennedy described his entrance as a time when "practically all of these [grants and contracts] were continuing and effectively noncompetitive. The next year's money for each project was almost automatically forthcoming in exchange for a short letter-form proposal outlining what the PI expected he might do during the next year or so and, if he happened to be in an expansive mood, perhaps a short summary of what he had been doing during the preceding years." While this system clearly benefited those researchers lucky enough to be funded, it did not forecast a robust future.

Into this stagnation Kennedy injected a tripartite vision of renewal. First, IIHR would pursue a diversity of new and innovative research initiatives. Second, these initiatives would address applied as well as basic engineering questions. And third, they would be funded by a mixture of government grants and industrial contracts. The unified vision challenged IIHR to consider a significant expansion of research fields and relationships. Although this chapter focuses on Kennedy and his successors as director, Kennedy's vision was enacted through the many skills and efforts of IIHR's research staff, that are delineated in the remaining chapters of this book.

Kennedy's tripartite vision was well rooted in practical considerations. He saw diverse, innovative research initiatives as the way—the *only* way—to maintain IIHR's very existence through the coming decades. "A continual stream of innovation," Kennedy wrote, "is essential to the continuing success of any hydraulic institute." Because many hydraulic problems had been addressed in the "golden age of fluids research (roughly 1890–1960) . . . , innovate we must because ideas are, after all, the sole product of [hydraulic] research institutes." Subject matter for innovation would pose no problem, for Kennedy recognized that the dire need for solutions to problems related to water use, river behavior, pollution control, and energy production were surfacing on the national consciousness.

For several reasons, this innovation would have to stem from applied research initiatives and their commercial connections. Primary among these was his belief that the "doers" (practicing engineers) and the "thinkers" (university researchers) played complementary roles, the former providing the latter with a continual set of questions and meaningful outlets for their creative energy. This productive alliance between the two "tends to keep basic research

John F. Kennedy, IIHR's third long-term director, steered IIHR toward industrially funded, applications-oriented grants and contracts. Promoting his vision of renewal and expansion, he wrote, "No research institute can long survive in a contemporary, competitive setting without a constant infusion of innovation from every level—model building and building maintenance, to theoretical analysis and development of patentable devices. The success of institutes that succeed innovatively is virtually assured."

from deteriorating into aimless research," Kennedy stated. Kennedy also saw university-based applied research as beneficial to students, with project connections affording students a relevant and marketable education as well as connections to the job market. And he viewed the practical application of one's knowledge as the hydraulician's duty and responsibility. "Hydraulicians must be ever mindful of the fact that their subject is one which is intended ultimately for application, and be cognizant of the needs of their constituency," he wrote. With time, duty evolved into a strength. In his 1991 resignation letter, Kennedy wrote that the staff's ability to "devise novel solutions to nettling problems" had become "one of the institute's hallmarks and chief drawing cards."

Kennedy also viewed applied problems as a vehicle for entering fundamental research efforts, rather than as a departure from them. "The distinction between basic and applied research lies primarily in the attitude of the researcher," he wrote. "Practically every applied problem has abundant basic-research challenge in it. . . . The good researcher recognizes and exploits this." Given time, this concept proved itself to be functional and correct. By 1990, new research lines routinely began as commercial projects and then were pursued with funding from NSF or other government agencies.

Funding for new applied initiatives quite naturally would have to flow from those who directly benefited from them—that is, their industrial sponsors. Not only was this reasonable, it was also necessary because of the cost of the technical innovations that were starting to emerge in the mid-1960s. Government grants were already becoming scantier. They would never be able to provide the funding for instrumentation, computers, and space to keep IIHR up-to-date.

The pursuit of applied initiatives became the legacy of Kennedy's quarter-century career, serving as the focus that he practiced at IIHR, fostered as president of the International Association for Hydraulic Research, and expounded in his professional papers. This pursuit redefined IIHR's mission. As commercial and business interests brought their problems to IIHR's doors, and these fed the imagination of IIHR's staff, stagnation was quelled. IIHR entered a period when growth in budget, staff, physical plant, and projects became the norm. These multiple forms of expansion furnish a testimonial to the successful, rich intermingling of today's applied and theoretical research efforts.

IIHR's budget soared from around $250,000 in the mid-1960s to $1.1 million in 1980, then resumed its steady climb to reach nearly $3 million in 1991, at the end of Kennedy's tenure. It then maintained a gradual growth, reaching $3.4 million in 1996. Much of the growth has originated from commercial projects, which were credited with rising from insignificance in the mid-1960s to funding about half of IIHR's annual budget by the early 1970s. Funding that had originated from only a dozen sources in 1967, nearly all of which were governmental, by 1991 came from over 40 diverse public and private sources. As new programs fused with and supplemented the old, total project numbers soared from 21 in 1967 to 81 in 1990 and 76 the following year. In the 1990s, it has remained about the same, fluctuating between 74 and 91 per year. Perhaps more important was the great diversification of both research initiatives and projects and of research approaches and techniques, a ballooning expansion that is described in following chapters.

The growth in research necessitated an expansion of facilities and staff. From 1975 to 1996, five research annexes have been constructed, increasing IIHR's floor space by 57,832 square feet. The electronic equipment housed in these buildings is tremendously more sophisticated than it was in the past, with computers having multiplied in role as well as number, in turn transforming the very type of research performed.

These buildings also house a far greater number of staff members, whose skills have propelled the innovations and expansion of the research program that distinguish the last few decades of the twentieth century. Senior staff members have increased threefold, from 9 in 1966 to 26 in 1991. In 1997, senior staff numbered 28, with 4 additional positions being advertised. The support staff has more than doubled, with the 9 shop and office workers in 1966 rising to 22 shop, office, administrative, and computer personnel in 1991, and dropping to 20 in 1997.

The total number of graduate students has not changed much. The 44 students of 1966 rose to 57 in 1991, but in 1997 dropped to only 39. In the past two decades, the number has fluctuated between 35 and 60 students. However, the involvement and breadth of their training has increased significantly. While fewer than half of all students were supported as research assistants in the late 1960s, by the mid-1990s virtually all graduate students received research or other stipends, and the vast majority were involved directly in an ongoing IIHR research project. As a result, today's students acquire a mixture of theoretical training and hands-on experience with actual engineering problems that serves them well in later jobs, and that spreads the legacy of IIHR's diversity and breadth.

Kennedy was born in 1933 in Farmington, New Mexico. He often talked about his rough-and-tumble childhood, being raised and educated on southwestern Indian reservations, where his father was an employee of the Bureau of Indian Affairs. Drawn away from New Mexico at age 18 by a concerned family friend who arranged for his admission to Notre Dame, Kennedy received a civil-engineering degree from that university in 1955, an M.S. degree in civil engineering from the California Institute of Technology (1956), and a Ph.D. in fluid mechanics and hydraulics from the same institution (1960). He spent the following year there as a post-doctoral fellow, and then moved on to an MIT teaching position. Rouse claimed that he drew Kennedy to Iowa by offering him "the [IIHR] directorship and a full professorship, which he accepted. . . . Such a windfall was not to be ignored." On July 1, 1966, Kennedy became IIHR's new director and Professor of Fluid Mechanics—a title that he ascribed in retrospect to the vogue of adopting scientific-sounding titles in the post-Sputnik decade of emphasis on science.

Kennedy brought several personal traits and beliefs to his position that promoted the success of his initiatives and thus shaped IIHR. For one thing, he loved its subject matter with a passion. "Our profession—the study and management of water—is indeed a fortunate one for several reasons," he wrote in a 1981 publication. He went on to enumerate the "great satisfaction on working in a non-destructive way with nature's creations" and the fact that "our activities are almost universally regarded as beneficial." His elemental attraction to the power and impact of water emerged from his early memories of a New Mexico childhood, he claimed. In that dry land, "I realized that without water, you had nothing—it is the key to utilization of the land, of people surviving." This human dependence on water was an asset to be both appreciated and capitalized upon. "Until we find a substitute for water or invent a way of preventing extreme weather events that cause flooding, there will be a big demand for hydraulic engineers," he quipped.

Kennedy's love of hydraulics remained strongest for his chosen specialty, the mechanics of alluvial rivers. He had written his dissertation on the fluid mechanics of dune and antidune formation in river channels. He returned to this subject time and time again, researching mechanisms of sediment transport, sediment management techniques, bedforms, and erosion control. "He loved every part of river research, from plodding along an eroded bank or inspecting bed armoring in freezing weather to formulating elegant

Kennedy, seen here with a model of an Iowa vane (an erosion-control structure that he helped develop), wrote of the passion and excitement he derived from being "engaged in a profession which permits one to function over the full spectrum of technology, from purely scientific to wholly applied."

solutions for river secondary flows and complex bed forms and guiding theses to completion," recorded an IIHR colleague.

With time, his love of hydraulics matured into a devotion to IIHR. A stubborn man who lived his work, he came to identify his own life and successes with those of IIHR. Writing late in his career of his decades-old obsession with setting IIHR off from other institutions and making it special, he revealed that he had "continually pondered what IIHR should then be doing, or should be doing differently, to assure that it would remain among the successful."

One more passion was his immense enjoyment of interacting with people, regardless of their background or professional interest. To most he was a charismatic man who knew how to attract others to him and make them feel comfortable and special. He relished energetic discourse and valued discussion with anyone with whom he could interact intelligently. He believed strongly in the exchange of ideas between practitioners and researchers, and his personal inclination led him to love interacting with practicing hydraulicians in the field, where he could come face-to-face with real-world, down-to-earth problems and with those who were struggling to solve them. Quick and witty in his responses, unpretentious in his approach, approachable and easy to those whom he knew and liked, he developed strong professional relationships with people and their agencies across the country. This mixture of love of profession, ideas, and human interaction fostered his extensive consulting activities and became a crucial component of his promotion of IIHR's applied-research program.

Another crucial component of applied research was Kennedy's killer-pace travels to projects and agencies around the world. They allowed him to keep his finger on the pulse of engineering applications. During these trips he nurtured connections with other professionals and ongoing efforts, established new contacts, and hustled projects and funding that he brought back to boost IIHR's ongoing workload.

Kennedy's finesse with both people and ideas was matched by yet another unusual combination: a rare joining of academic and capitalistic prowess. While solidly executing his professorial duties, he simultaneously read books and magazines on business practice and skillfully applied their suggestions to the management of IIHR.

His interest in business was complemented by a seemingly innate ability to sense future directions and to develop his intuitions about future needs into viable research programs. Quick of mind and temperament, he identified emerging research areas early enough for IIHR's researchers to claim the initial funding and publish the initial papers in several areas. He recognized himself as a creative innovator, an "ideas person," an impatient man who was always chafing to move in new directions and get on with things. He thrived on the excitement of a complex problem with its messy interweaving of technical issues and politics. While quick to grasp the salient features of an engineering problem and conceive possible solutions, he was also quick to tire of a subject as its explorations became more routine. He then mentored younger researchers in the given subject, drawing back and participating from a distance, while focusing his own investigations on something new. Thus, he established himself internationally in several areas in addition to riverine processes: ice engineering, cooling-tower design, dropshafts for urban storm drainage, waste heat and its disposal, as well as density-stratified flows and the turbulent mixing of fluids. Perhaps more importantly, he established these fields as major IIHR research foci and funding sources and greatly aided junior researchers in founding their careers in these fields.

Kennedy's inquisitive nature and joy in the new led him simultaneously in multiple other directions as well. Like Rouse, Kennedy lived hydraulics and put work at the top of his life's priorities. However, while conversation with Rouse always circled back to hydraulics, talks with Kennedy might take the listener anywhere. He was an avid reader of diverse books and loved discussing his most recent find. He learned as much as possible about music, literature, history, sports, and a myriad other topics that interested him. With his prodigious memory, he had plenty of fuel for energetically debating numerous topics, an activity that he relished. His neighbors learned that he might drop by not only to discuss the fluid mechanics of the garden hose or the formation of icicles, but also to share his knowledge on any of many subjects. He was well known for quoting show tunes, in particular those of Cole Porter and Gilbert and Sullivan.

Kennedy made decisions rapidly, following his gut-level instincts with assurance. Commended for his strong leadership and organizational skills, he led others by his own example of immersing himself fully in his work. An independent man who didn't like others to tell him what to do, he allowed IIHR's Board of Consultants to fade into oblivion. This traditional advisory committee of prominent outside engineers had first been configured by Floyd Nagler in 1931. Kennedy duly appointed new members for the first few years after his arrival. He listed them, as was tradition, in his annual report until 1970, when he simply mentioned the outgoing members and failed to appoint replacements. Thereafter, the advisory committee was not mentioned or convened.

These many traits were fueled by a personal energy and vitality and by an intense competitive spirit, both of which were uncommonly strong in Kennedy. Driven to a frantic pace of living, he did everything hard, be it work or play. Vigorous, aggressive, dynamic, outspoken, with a flair for the dramatic—all these traits fed into Kennedy's shaping of IIHR's research program. His sense of competition fueled his obsessive personal drive. He constantly challenged himself in complex problems as well as simple matters that added humor and excitement to the day—for example, by racing up IIHR's enclosed stairwells, trying to make it to the top of a flight before the bottom door swung shut. As his long-term administrative assistant, Marlene Janssen, wrote:

> He was never hard to find. He was the one in the center of the group that was laughing most heartily; or the one involved in the most spirited debate. He was the one walking the fastest or being approached most often by young engineers, who were never slighted by him. He was the one presenting the most thought-provoking lecture or delivering the most entertaining banquet speech.

His energy spilled into numerous traditional professional efforts. He took pride, for example, in IIHR's international exchanges and had a keen interest in IIHR's effect on hydraulics around the world. In addition to consulting and spending sabbaticals abroad, in 1978 he led the official delegation of U.S. river engineers on a technical tour of the People's Republic of China, and in 1982 and later years he helped review development of a national water plan for Saudi Arabia. In 1985 he planned a 16-day tour for himself and the university president to the People's Republic of China, in order to promote student and faculty exchanges with that country. His Chinese involvement led to his advising on sediment management and navigational aspects of China's Three Gorges Dam, still under construction, which when completed will be the world's largest hydroelectric plant.

Kennedy was a prolific writer. He authored or co-authored over 250 publications. The vast majority of these were technical papers dealing with specific research questions. Some address the value of pulling commercial interests into university research laboratories. He also enjoyed writing review talks and papers. Toward the end of his life, he (like Rouse) focused on the history of hydraulics, and he composed several biographical sketches of hydraulicians of the twentieth century.

While these multiple efforts and traits of a strong-willed man put IIHR's research program on the fast track, Kennedy's activities were not ubiquitously praised or viewed as positive. His hectic travel schedule kept him on the road and limited his personal time in the classroom. The educational innovation that had characterized

Rouse's years withered. In addition, Kennedy was a stubborn man who strictly apportioned building maintenance responsibilities to the university and educational responsibilities to the college. Thus, renovation of IIHR's older buildings and modernization of its teaching laboratories took a back seat to investments in new research facilities and equipment. He also refrained from hiring new personnel, preferring to assign new tasks to already overworked staff members who became progressively more overburdened. Even his energetic and innovative fostering of applied and contract work was criticized by a few, who saw it as good business but also as a nonscholarly emphasis inappropriate for an academician and detrimental to IIHR.

Kennedy's personal style also could be abrasive, his strong leadership domineering and controlling rather than democratic. His constant competitive spirit and need to be on top of a situation stimulated snappy comebacks that sometimes could be humorous but at other times could be cutting, cruel, and vengeful. He elicited strong responses from others, and while he maintained the loyalty of most of IIHR's staff, some viewed him negatively. Yet with his witty humor and constant jokes, his ability to laugh at himself, his warmth, and his genuine caring for others, Kennedy for the most part elicited an affection and a loyalty from staff members that fared IIHR and its initiatives well.

Kennedy was well honored and recognized for his talents and years of devotion. Elected to the National Academy of Engineering at age 39, he became one of the youngest men ever to join its ranks. He held first a Carver Distinguished Professorship at Iowa and later the Hunter Rouse Chair in Hydraulics. He received Fulbright and Erskine fellowships. Several foreign associations bestowed honorary memberships on him, and ASCE presented him with four awards for his research and publications. His expertise was sought on national and international consulting boards. Kennedy regularly hobnobbed with the top administrators of the University of Iowa, served on several review committees and research committees for university vice presidents and presidents, and even applied once himself for the presidency of the University of Iowa. He also served from 1974 to 1976 as chair of the Engineering College's Energy Engineering Division, and as two-term president of the International Association for Hydraulic Research, in that position winning praise for the discussion and integration he encouraged between applications-oriented hydraulicians and those focusing on basic research.

People could not resist kidding John F. Kennedy, hydraulician, about his relationship to John F. Kennedy, U.S. president. (Once, tiring of the jokes, he responded that his mother had named him after New York's John F. Kennedy airport.) However, the comparison was not totally absurd. Both Kennedys assumed their leadership positions at an amazingly young age and both carried out their duties with vigorous insight, boldly calling for new dedication to the tasks at hand. Both died before their time. In mid-July 1991, Kennedy wrote to tell Hunter Rouse that in 1987, he had been diagnosed with multiple myeloma, an incurable bone-marrow cancer. "In the 4½ years since then, I just have continued with life—with drug therapy added to it," he wrote. Keeping his illness from all but a few of his closest associates, he had kept up his travels, teaching, writing, research, administration, and consulting, working more intensely, if anything, than before, burning the proverbial candle at both ends. However, his zesty continuation of life's activities could not prevent the cancer's progress. Three weeks before writing to Rouse, he had announced his resignation as director of the institute, to be effective at the end of August, a quarter century plus two months after he had come here. Not mentioning illness as a cause, he wrote to the dean, "I find the relentless company of the administrative duties of the Director to be increasingly burdensome. It is time to relieve myself of them, lest I forget completely how to be an engineer and a researcher and a teacher. . . . I take this step with great ambivalence for, as you know, IIHR is in the very marrow of my bones. . . . I could not have had a more supportive and happier setting for the pursuit of my career." Though his letter implied his intent to continue activities at the institute, his health continued to fail, and death overtook him on December 13, 1991, four days before his 58th birthday.

Kennedy's apparently rapid decline and death shocked many within and outside IIHR's walls. As when Floyd Nagler had died 60 years earlier, IIHR was suddenly devoid of a strong leader who most had assumed would remain in charge for many more years. However, while Nagler's death left the college scrambling for administrative leaders and funding, Kennedy's death was followed by a smooth transition period that was adeptly guided by Robert Ettema.

Ettema had come to IIHR in 1980 as a post-doctoral associate, straight from the University of Auckland, New Zealand, where he had received his undergraduate and graduate degrees in civil engineering. He had joined the faculty a few years after his arrival in Iowa and had proceeded to establish himself as an ice engineer and river hydraulician. In 1986, Kennedy had asked Ettema to serve as associate director of IIHR in order to relieve Kennedy of some of his administrative duties and allow more time for pursuit of his own research. In September 1991, immediately following Kennedy's resignation, Ettema assumed the position of acting director, and a month later IIHR research scientist Tatsuaki Nakato became acting associate director. These two administered IIHR for three years while recruitment of a new director was stymied by an extended vacancy in the Engineering College dean's office, due to the retirement of Dean Hering and the death of his interim replacement Paul Scholz.

Although an acting director, Ettema was by no means inactive. One of his major accomplishments was maintaining the spirit of staff members and holding an even keel in potentially turbulent waters. Kennedy had initiated many ongoing research initiatives and then had maintained close relationships with the projects and their funders. His sudden disappearance might have caused a sudden loss of confidence of both staff members and those who were providing projects and funding. This did not occur: research projects and funding remained healthy, firmly grounded in the skills of a good-sized, diverse, and deeply committed staff. Thus, IIHR, which through its 70-plus-year history had been headed primarily by three strong and directive leaders, was shown to have matured into a stable, resilient institution no longer dependent upon, or defined by, a single person.

Like other new directors before him, Ettema attempted to redress the omissions of his predecessor. Although Kennedy had presided over a period of tremendous growth of projects, laboratory space, and equipment, he had not invested in IIHR's infrastructure. IIHR's facilities had become shabby. They no longer invoked a sense of professional excellence. In addition, IIHR's need for office space and computer laboratories had soared. Ettema took IIHR's physical deterioration to heart and commenced a period of renovation, departing

University Relations

Robert Ettema, Associate Director of IIHR (1986–91) and Acting Director (1991–94).

from Kennedy's example by unabashedly using IIHR's funds to do so. The Hydraulics Laboratory's fifth floor was converted into good-sized faculty offices, IIHR's classroom was moved to the first floor, the flood-prone library moved to a drier location, and the mechanical shop was consolidated into one physical unit located in a new expansion of the East Annex. Computer space on the third floor was expanded, and the Computational Laboratory for Hydrometeorology and Water Resources was moved here from the Engineering Building. Ettema planned that additional much-needed offices would be constructed in a mezzanine of the Wind Tunnel Annex.

Finally on June 1, 1994, IIHR's seventh director was appointed. Virendra C. Patel had received an undergraduate degree in aeronautics from the Imperial College of London, a doctorate in the same from Cambridge University in 1965, and had come to IIHR via the Lockheed aerospace firm in Georgia as an assistant professor in 1971. He had been instrumental in introducing numerical-modeling techniques to IIHR and had been applauded and awarded both for his teaching skills and for his research in boundary-layer theory, tur-

Tatsuaki Nakato, Acting Associate Director of IIHR (1991–94) and Associate Director since 1994.

bulent shear flows, wind engineering, and ship hydrodynamics. His faculty affiliation with mechanical engineering once again melded the inclinations of IIHR's director with the needs of the times—that is, with the attempts of multiple researchers to use computer technologies to better understand fluid flows. Patel retained Nakato as his associate director.

Before V. C. Patel accepted his new position, he had successfully negotiated three new faculty positions and $1.2 million of college and university funding (to be matched by $1.0 million from IIHR) for additional renovations and equipment purchases over five years. Soon after stepping into his office, he initiated a comprehensive master plan for a stepwise renovation of facilities. Soon thereafter, the Wind Tunnel Annex mezzanine offices were constructed. Additional offices were constructed on the third floor, Hydraulics Laboratory, further moving this structure toward becoming the office and computer headquarters for the multiple outlying laboratory annexes. The entire Hydraulics Laboratory was rewired for safety purposes and to provide for increased power demands, its

Virendra C. Patel, IIHR's Director since 1994.

old steel water supply lines were replaced with copper pipes that could provide adequate pressure, and the leaking roof was re-covered. In 1996 IIHR's fifth annex, Oakdale 2, was constructed.

Patel started taking a much broader interest than Kennedy in en-couraging and financially supporting the library and IIHR's educa-tional initiatives. For example, he initiated the planning of a new state-of-the-art fluids teaching laboratory that will be utilized by college undergraduates and graduate students, and he instigated the International Perspectives in Water Resources Planning study-abroad program. He also helped move IIHR's finances to a more businesslike foundation by hiring IIHR's first professional accoun-tant, Paul Below, in October 1994.

Patel seems to be taking a gentler and more relaxed view of his leadership role than his predecessors. He is neither pronouncing certain endeavors as taboo, as Rouse did, nor out hustling research contracts as Kennedy did. Rather than imposing a firm and possibly inhibitory structure from the top, he is encouraging IIHR's staff to identify and initiate their own research programs and thus ulti-mately allowing them to determine IIHR's future direction. His actions center around supporting staff members in doing so through actions such as purchasing IIHR's new supercomputer and revising IIHR's fiscal policy. While in the past research engineers could only be hired if funds were available from ongoing grants and contracts, and then were required to bring in a major portion of their salary from these sources, researchers in 1997 have their salaries assured from college funding, with the expectation that they will continue to perform at the same levels as they have done in the past. Patel feels that this relieves considerable strain that in the past has forced some researchers to leave, and also encourages the mentoring and development of junior research scientists who are just establishing themselves. In keeping with his more democratic approach to the directorship, he accepted this position on a four-year-term basis, to be renewed only after review and input from the staff and dean.

Looking back over those who preceded him, Patel has many examples to follow: Floyd Nagler with his passion for a bigger and better Hydraulics Laboratory structure and his love of river research and its applications, the directors who struggled to maintain an institute and research program during the Great Depression, Hunter Rouse, who fixated on applying fundamental fluid mechanics to hydraulics through teaching, research, and writing, and John Kennedy and his attempt to rebuild a firmer funding base and more balanced research program through resurrecting IIHR's applied-research initiatives. Patel's legacy as director is only just starting to unfold. However, with the interest that he has demonstrated in education, applied and theoretical research, and IIHR's physical structure, he may one day be seen as the director who took a healthy IIHR into the twenty-first century by combining the best of those who had gone before.

13. Research, 1966–97

Toward the end of Hunter Rouse's directorship, IIHR was in danger of losing its status as a premier U.S. hydraulics laboratory. While its contributions to basic research and education were prodigious, its fiscal condition was becoming precarious. Staff numbers remained small, their research funded by a few agencies and grants. Lack of growth in staff size, funding, funding sources, and project diversity created a sense of stagnation and uncertainty.

The impending threats within IIHR mirrored changes occurring across the country. The era of broad-based long-term funding, doled out without strings to productive researchers rather than to specific goal-oriented projects, was drawing to a close. This was the "sunset era" of the large, generous water-reclamation, hydropower, and navigation projects that had dominated aspects of U.S. hydraulics for many years. At the same time that federal financial support was shrinking, increasingly sophisticated and expensive instruments, including computers, were rapidly becoming research necessities. If the institute was to maintain its stature—if it was to survive at all—it would need to chart a new course that diversified its research program and provided ample funds.

Future possibilities depend in part on one's vision. "So-called routine model testing affords abundant opportunity for innovation if the engineer doing the tests is of the innovative sort." Thus wrote John Kennedy, IIHR's new director, dismissing with a stroke of the pen Rouse's proscription of model-based and site-specific applied research. In his first annual report, Kennedy pointed out more specific possibilities for the future:

On every side there is overwhelming evidence of an acute and still growing national awareness of technical problems relating to man's environment and health, problems into which the discipline of fluid mechanics enters as a central ingredient. . . . Moreover, there is a growing body of evidence that these and similar [environmental] problems may soon be receiving the major research and development thrust of the nation . . . and that these problem areas will continue to be regarded as . . . important for many years to come. Accordingly, it is both logical and prudent for the Institute to orient some of its activities in these directions.

With such statements, Kennedy displayed one of his stellar traits—his uncanny ability to predict future directions in research funding well in advance of their appearance. During the next few years, the federal government and the general public did indeed become increasingly aware of major dilemmas in the natural environment that significantly impacted human health and society's future. Many of these involved the distribution, protection, and use of water. A new environmental consciousness pointed to the need for better management of water resources, environmental protection, and reclamation. With the signing of the National Environmental Policy Act (NEPA) in 1970, these concerns began to work their way into legislation. Shortly thereafter, in the early 1970s, the first energy crisis led to a boom in power-plant construction. All these events represented research opportunities to the hydraulic engineer, who was being called into a new covenant with society, a covenant that factored the social and environmental consequences of technological developments into professional decisions.

With a rare mixture of academic and capitalistic prowess, and backed by a creative, competent staff, Kennedy led IIHR into exploring emerging research initiatives and funding sources. "Without dropping old subjects, the Institute is swiftly moving into new subjects," proclaimed IIHR's Christmas letter for 1967. Four years later, it announced that "in keeping with the winds of change in the world of science, the Institute is striving to reorient its priorities. . . . A little butter and much of our bread comes increasingly from our more recently undertaken research projects." For the next two decades, Kennedy—the third long-term director who stubbornly re-created IIHR in the image of his own vision—nudged and whittled IIHR into a center for applied, project-based research, much of which focused on the broad interface between human action and the natural environment. This emphasis became a yeasty brew when blended with the projects Rouse had fostered—the fundamental-fluid-mechanics studies of wakes, jets, and turbulence; traditional hydraulics and hydrology; instrument development; and the early development of numerical-modeling techniques.

The result was growth on all fronts: growth of IIHR's staff and students, physical plant, instrumentation and equipment, budget, number and size of research projects, their funders, their diversity. The bulk of the funding for this growth came from commercial projects, which by the early 1970s had expanded to provide about half of IIHR's annual income, while the remainder still flowed from basic-research projects. Thus, IIHR thrived during a period when less adaptable hydraulic-research laboratories were going belly-up, for although the responsive labs flourished, there were far fewer hydraulics laboratories in the 1990s than in the 1960s. A vital alliance with industry profited IIHR sufficiently that when looking back on the mid-1960s through the 1980s, Kennedy dubbed this interval the "golden period for hydraulic institutes."

Much of the remainder of this book examines the resulting rich diversity of research that still characterizes IIHR. Chapters 15 through 20 attempt to create a sense of the vitality and excitement that distinguish many projects, and to illustrate their contributions to the workings of the everyday world. These chapters focus in detail on a single aspect of modern research.

The present chapter paints a thorough chronological mural of IIHR's research with broader brushstrokes, emphasizing subjects not described in later chapters. It elucidates changes in research emphasis and focus by describing the historical roots and development of many project areas, as well as some subject areas that were important in past years but are no longer emphasized. This chapter, in unison with the later chapters, portrays IIHR's research during the 30 year period 1966–1997 in its entirety.

Power-plant Research

Soon after Kennedy's arrival, grants were received for two new studies. The first, looking at air flow through large cooling towers for thermal power plants, was funded in 1967 by a manufacturer of these cooling towers, the Marley Company. The second, a study of dispersion of water pollutants in winding streams, was funded in 1968 by the Federal Water Pollution Control Administration. These studies were unlike anything that IIHR had done in the past. However, although they were harbingers of things to come, it remained the task of a hotly debated nuclear-power plant to fully pull IIHR into a new environmental-research focus concerning the multiple facets of steam-electric power plants and their thermal effluents.

Construction on Commonwealth Edison's Quad Cities Nuclear Power Plant started in 1966. At that time, the nation's demand for electricity was doubling every ten years. Similar continued increases were assumed as inevitable, and "living better electrically" was advertised as the means for achieving the American dream. Over 80% of the nation's electrical energy was generated by thermal (steam-electric) power plants that burned either fossil or nuclear fuel, and that percentage was increasing. A major problem with such plants was the generation of waste heat, an inevitable by-product of an inefficient process: much of the energy from steam can be transformed into electricity, but about two-thirds of the energy is given off as heat, which must be removed if the plant is to work efficiently. The trend then was toward more and larger nuclear power plants, which produce 50% more thermal water pollution per unit of power than do fossil fuel-burning plants.

Most power plants at that time got rid of their waste heat through "once-through" cooling systems. These systems, which were much simpler and cheaper than cooling systems that recirculated the water, would draw water in from a river or lake, feed the water through condensers to capture the waste heat, and dump the heated water back into the river or lake, heating the natural water body perhaps several degrees. Little thought was given to the consequences of this common practice. Concern about water pollution was just beginning to work its way from a growing national awareness into federal legislation. The first Federal Water Pollution Control Act had been passed in 1948, but early legislation focused rather narrowly on human health and the construction of municipal waste-treatment facilities. The establishment of the Federal Water Pollution Control Administration in 1965 called for water-quality standards and the development of implementation plans for cleaning the nation's waterways. By 1970, the newly formed President's Council on Environmental Quality was identifying waste heat as "one of the most serious emerging sources of water pollution." That year, the year of NEPA, controls for interstate pollution were tightened. Finally, in 1972, amendments to the Water Pollution Control Act empowered the Environmental Protection Agency (then just two years old) to establish regulatory guidelines for a number of water qualities, among them the temperature of power-plant effluents.

The Quad Cities power plant initially was fitted with a once-through cooling system. However, by 1970, when the plant was nearing completion, such cooling systems were being examined more critically. IIHR thus was contracted to perform a model study of the plant's thermal outfall. Researchers learned that the many thousands of expected gallons of cooling water would be heated by as much as 23°F by the plant and discharged as a single jet. The jet would eventually disperse, but even three-fourths of a mile below the discharge point, the heated water would raise the river's temperature 3° or 4°. Wildlife agencies expressed strong concerns about the ecological effects of such heating. IIHR was asked to come up with an alternative water-discharge system. It designed a pair of perforated pipes, each 16 feet in diameter, that would stretch 2,000 feet across the Mississippi. The pipes would be buried in the riverbed and discharge heated water through perforations every 20 to 40 feet along their length. The discharging heated water would raise the river's temperature less than 5° at any one spot, thus staying within federal guidelines. However, controversy about the ecological effects of such a solid "wall" of heated water continued until 1974. Newspaper debates accompanied multiple regulatory hearings, and Commonwealth Edison was once again denied permission to operate the plant. Eventually a compromise was reached: the plant could install and operate the discharge pipes during the time required to implement a closed-circuit system—one that circulated the cooling water through a cooling tower, cooling pond, or spray system, and then reused it.

Through the controversy, many IIHR researchers were pulled into field, model, design, and monitoring studies of the Quad Cities discharge system. These studies continued until 1991. They opened the door to many additional projects as old questions were answered and new ones were asked—a research-expanding process that came to characterize many of IIHR's applied projects in these years. For example, IIHR considered the assets of waste heat. Could it be used to clear ice from portions of the upper Mississippi in winter and thus keep its navigation channel open? Kennedy foresaw the possibility of locating multiple nuclear power plants at intervals designed to provide the correct amount of ice-melting heat. Consequently, grants to fund general studies of thermal effluents' effects on river ice were received from 1971 through 1976, fusing IIHR's cold-regions and power-plant efforts. With concerns about the ecological effects of Quad Cities' thermal effluents rising to the fore, Donald McDonald, an aquatic biologist in the Civil Engineering department, joined IIHR's staff in 1972. Soon, biological-monitoring studies appeared on IIHR's research roster, typically in conjunction with power-plant projects. While investigating the effects of the Quad Cities power plant's effluents, McDonald identified a more obvious problem: the plant's cooling-water-intake structures were sucking in and killing large numbers of fish. Thus, IIHR commenced model studies of intake structures in order to reduce the entrainment and impingement of aquatic organisms.

These applied studies complemented ongoing basic research in heat transfer and dispersion of buoyant pollutants in free-surface flow. Fundamental research projects on dispersion of thermal pollutants and on cooling-tower operation jumped from one to six in 1970 and continued to increase thereafter. The Quad Cities disputes, in combination with stricter federal regulations for thermal pollution, also drew IIHR into a steadily increasing flow of other applied studies funded by a diversity of private sources. Within five years, IIHR was conducting model studies of power plants and their discharge systems from Hawaii to Florida. Many involved structures for pulling water into power plants or releasing thermal effluents into natural water bodies or cooling ponds. Others included the mid-1970s MAPP project, a mapping of the thermal regimes of the Mississippi and Missouri drainages, performed with an eye toward planning the siting and number of future power plants with their resulting thermal effluents.

Still other research dealt with cooling towers—structures that dump waste heat directly into the atmosphere. The large waste heat outputs of nuclear power plants meant that most had to be

Power-plant research dominated IIHR in the 1970s. Studies of the Jack Watson Station in Alabama (above) were sufficiently complex to require the construction of IIHR's mammoth Environmental Flow Facility in the mid-1970s to investigate the performance of a unique, circular cooling tower for this plant (below). This remains IIHR's largest hydraulic flume.

retrofitted with closed-circuit systems to meet the new federal thermal-pollution guidelines. In 1974, IIHR commenced a study of the economics of such retrofitting and compared the cost of cooling towers, cooling ponds, and similar systems. Another study compared the economic and water-use efficiency of various cooling systems. Later investigations sought ways to increase this efficiency. Meanwhile, IIHR continued model studies and fundamental research on the basic operation of cooling towers—their aerodynamic loading, dispersion of heated air, resistance to wind forces, most efficient geometrical arrangement, recirculation of plumes, etc.

Power-plant and related thermal-diffusion studies quickly became major supporters of IIHR activities. The number of power-related projects soared from seven in 1970 to a dozen in 1972 and 25 the following year—over half of IIHR's projects in 1973. The proportion held at around 50% through 1976, dropping to around 40% in the late 1970s but still numbering one to two dozen projects a year. In the early 1980s, power-plant construction across the nation declined for several reasons: an excess generating capacity resulting from the 1970s construction boom, a switch in emphasis to energy conservation, and more stringent governmental regulations on new constructions. The number of IIHR's power plant-related projects dropped 90% in response.

However, power plant–related studies did not totally disappear. Since 1980, IIHR has continued to receive requests for help with managing and monitoring existing plants. It has investigated the continued operation of cooling towers, cooling ponds, and cooling-water intakes, problems with sediment blockage and discharge systems, and impacts of thermal effluents. Other types of power-related studies have been undertaken, for example investigations of the potential benefits of hydropower plants on Iowa rivers, reservoirs, and the Mississippi River. McDonald's environmental-monitoring studies expanded in 1977 to include water-quality monitoring of the Coralville Reservoir, an effort continued by research scientist Kent Johnson. In 1983, IIHR was asked to investigate salmon-diversion structures for the large hydroelectric plants and dams that spanned the Columbia River in the Pacific Northwest. The focus was fish-blockage problems similar to those that IIHR had investigated for small Iowa dams in the late 1930s, and fish-impingement problems such as those McDonald had discovered a decade earlier. The salmon efforts grew while other power plant–related research continued to shrink. Carried on by Jacob Odgaard and Larry Weber, who joined the research staff in 1977 and 1996 respectively, these studies constitute the single largest portion of IIHR's funding and provide a major research focus, forming a continuing legacy to IIHR's early power-plant efforts.

Ice Research

Ice engineering became IIHR's second major research thrust of the 1970s. Like the power-plant studies, it was shepherded into IIHR's folds by Kennedy. Although ice studies remained less numerous than power-plant studies, ice together with power production dominated the research agenda of that decade.

Countries with cold-water shipping ports such as the Soviet Union, Japan, and the Scandinavian nations had long been interested in the subject. However, the U.S. had lagged behind until the early to mid-1960s, when interest had been spiked by development projects in Alaska and ice problems in the Great Lakes and Mississippi River drainage. Within months of Kennedy's arrival in Iowa, he announced that he was considering establishing ice studies as one of IIHR's major research initiatives. He solicited funding to construct a low-temperature laboratory, the first such university-owned research facility of its type in the Western Hemisphere, with comparable structures only in the Soviet Union and Czechoslovakia. By the time it opened in early 1970, the U.S. Army Corps of Engineers (COE) was supporting winter field studies of ice on the Iowa and Cedar Rivers. These were collecting prototype data for calibrating the laboratory ice that would be made in the refrigerated flume, and were the first phase of a continuing COE contract for addressing ice suppression and the possibility of maintaining year-round navigation on northern rivers. Three additional grants were received within months, to study the formation of ripples on the undersurface of river ice, river ice jams, and hydrologic responses of ice-covered streams. Resulting papers on ice ripples (by G. D. Ashton and Kennedy) and ice jams (by M. S. Uzuner and Kennedy) later won ASCE's Karl Emil Hilgard Hydraulic Prize in 1974 and 1978 respectively.

A major IIHR program in ice research was under way, one of the first which grew to be about as large and broad as any in the nation.

John Kennedy (foreground) launched the design of a "unique low-temperature flow facility for research on . . . river and lake ice" a year after his arrival at IIHR. Soon thereafter, NSF and university funding was received for constructing a recirculating flume with chilled liquid, housed in a refrigerated room on the second floor of the Hydraulics Laboratory.

Since the 1970s, numerous topics have been explored, with between five and ten projects usually proceeding simultaneously. In 1985, the volume of ice research kept the ice labs booked for two years in advance. The following year, IIHR hosted the IAHR's Eighth International Symposium on Ice, only the second time this prestigious conference was held in this country and an event that crowned

IIHR's ice facilities as some of the world's most active. Efforts in ice research regularly merged with efforts in other fields, each project multiplying into new questions and studies. In the mid-1980s, for example, IIHR's ship studies expanded to examine the motion of ship hulls plowing through ice sheets and rubble ice. Sediment transport under ice also garnished several grants over the years. Some of the ice projects are described in chapter 17, as are the several laboratory facilities that have replaced the original ice flume.

Several IIHR researchers were involved in these studies, with Kennedy maintaining a leadership role and J. C. Tatinclaux playing an active role from shortly after his arrival on staff in 1973 until 1982. Robert Ettema then took the lead, two years after his arrival at IIHR, when Tatinclaux left for a position at the U.S. Army Cold Regions Research and Engineering Laboratory. Wilfrid Nixon, whose research effort is devoted to ice problems, came to IIHR in 1988. The focus of research has evolved with changing times and personnel, with questions related to northern regions declining after oil prices (and northern oil exploration) dropped in the mid-1980s, river ice studies continuing but at a lower level than in the past, and road ice studies picking up the slack.

Biomechanics

A third focus initiated with Kennedy's arrival was biomechanics, the study of fluid mechanics of living systems. In 1967, research engineer Enzo Macagno and Art Giaquinta, a graduate student who later joined IIHR's staff, received an NIH grant to model red blood cells to determine stresses placed upon them in the blood stream. This grant opened the door for another NIH project the following year, performed jointly with the College of Medicine's Urology department, concerning the hydrodynamics of urination in young girls. One year later, Macagno received yet another NIH grant jointly with an Internal Medicine professor, to examine the mechanics of segmentational pumping in the small intestine. They considered how fluids could be pumped through the intestines by apparently random muscular contractions. The study involved volunteers who swallowed tiny pressure gages that recorded their intestinal contractions, as well as the construction of a model intestine: a 2-foot-long tube of soft plastic that could contract at six points, pushing forward its recirculating dyed water.

While these studies seemed for a time to constitute a major new initiative, and while the intestinal contraction study continued uninterrupted until 1983, biomechanics never became a primary IIHR research focus. Three more studies of blood flow through the aorta and heart were performed in the 1970s and early 1980s. In 1979, IIHR contributed to an attempt to designate the university as an Olympic training center for swimmers. The idea was to modify the very large environmental flow facility so that its circulating water could be used for physiological testing of swimmers. While Iowa never won the coveted Olympics designation, IIHR's flumes were used to train and test racing swimmers, and IIHR was involved in designing an open-topped water tunnel for swimming research in New Zealand. More recently V. C. Patel, in collaboration with F. Alipour-Haghighi of Speech Pathology, performed experiments and numerically modeled the flow of air in the larynx. Apart from these occasional efforts, biomechanics has remained an anomaly.

Sediment Transport and Hydraulic Modeling

River studies associated with sediment transport and hydraulic structures have been ongoing at IIHR for decades. These topics were for many years the "bread and butter" of the institute. Sediment-transport research commenced at IIHR in the 1930s, with fundamental and applied sediment studies expanding to form a major focus from the 1940s to the late 1950s. Following a hiatus in sediment research caused by lack of funding, studies recommenced in the late 1960s as a correlate of environmental engineering.

Starting with a sediment study here and there, by 1977 the IIHR holiday letter was proclaiming that this "traditional staple of hydraulic engineering is swiftly returning to the forefront of IIHR activity." That year, related projects included investigations of sedimentation in reservoirs, the sediment regime of a specified watershed, scour around bridge piers, and the effects of riverbed degradation on power-plant intake systems. Bill Sayre (a research engineer specializing in transport processes of sediment and water contaminants, on staff from 1968 to 1980) organized two workshops that sought the cause of continuing channel degradation of the Missouri and Des Moines Rivers. From then onward IIHR has feasted on a regular diet of investigations aimed at decreasing the need for dredging rivers and reservoirs and creating computational models

Studies of sediment transport, carried out sporadically throughout IIHR's history, were revitalized by concerns about the natural environment. "We're facing more environmental problems," wrote John Kennedy about this research topic. "Everyone wants the water, but no one wants the sediment." Here a student investigating sediment prepares to drop a current meter into the Mississippi River.

for river flow and sediment routing. IIHR has worked to explain why sediment consistently built up around certain Mississippi River islands, the mechanisms of streambank erosion and sediment transport around river bends, sediment routing during floods, and the sediment budget of the Upper Mississippi. Although carried out by a number of IIHR researchers, much of this work has been completed by Subhash Jain and Tatsuaki Nakato, river hydraulicians who joined IIHR's staff in 1971 and 1975 respectively and have undertaken a variety of sediment-transport and applied model studies in the ensuing years. Adsorption of pollutants to sediment particles was looked at by Jerald Schnoor, a research engineer who joined the staff in 1978.

Some of these studies concerned hydraulic structures designed to abate sediment problems, for example vanes for reducing streambank erosion (discussed more below) and pump-intake structures designed to minimize problems with both sediment and vortices. Model studies of hydraulic structures, which had been curtailed during the Rouse years but had never completely disappeared, now rose in prominence. Many of these, such as efforts to design dropshaft chutes for underwater storage systems in large cities (see chapter 16), were applied efforts tied to environmental problems.

Hydrometeorology

Research relating to hydrology, an emphasis dating back to IIHR's earliest beginnings, was revived and transformed in the 1980s. One of Floyd Nagler's first projects had been the establishment of monitoring stations for rainfall and runoff throughout the Ralston Creek drainage just east of Iowa City. Between 1974 and 1977, Tom Croley, a water resources specialist who had arrived in 1972, used the Ralston Creek data for a study of the hydrological impact of urbanization, for by that time the agricultural character of the watershed was being lost as Iowa City expanded eastward. He and Jain then used that work to construct a computerized model of flooding and sedimentation patterns in an urbanizing watershed—the very first numerical model created at IIHR and a precursor to the extensive computational hydrological, hydraulic, and fluid-mechanics models of today. Croley left IIHR in 1980. A second water-resources specialist and hydrologist, Peter Kitanidis, had come the preceding year and stayed until 1984. Limited opportunities for funding in hydrometeorology caused Croley and Kitanidis to be negatively affected by the requirement that IIHR's researchers bring in a large portion of their salaries from outside grants, an expectation that remained true until the mid-1990s.

In 1988, the U.S. Department of Agriculture support for the Ralston Creek project ended and a USDA representative came to remove the rain gages, an act that signaled the end of IIHR's traditional hydrology program and longest continuous effort. By that time, the institute's weather research was becoming far more involved and its equipment was far more sophisticated. A flurry of new hydrometeorologists arrived in the 1980s—Konstantine Georgakakos (on IIHR's staff 1985–94), Witold Krajewski (who came in 1987), and Ignacio Rodriguez-Iturbe (here 1989–91). Support for their efforts took off almost immediately, rapidly soaring to eight or ten grants per year. In 1990, they established a weather station on the roof of the Hydraulics Laboratory, which has since been moved to the East Annex's roof. Their ultimate goal was to develop techniques for remotely determining precipitation with great accuracy and for forecasting extreme events so that their destructive forces could be minimized. Their efforts have evolved into the hydrometeorology and the integrated watershed studies described in chapter 18, which have been carried on by Krajewski, Frank Weirich (here since 1988), Allen Bradley (since 1994), Anton Kruger (who became a research scientist in 1996), and Bill Eichinger (who came in 1997).

Instrument Development

Another major thrust, instrument and equipment development, had decades earlier been transformed into a research category of its own by Rouse's implementation of fluid-mechanics teaching laboratories and IIHR's need to convert from manual to electronic instruments. The initiative suffered a blow when Philip Hubbard, who had directed such efforts, accepted the position of dean of Academic Affairs in 1966. However, similar efforts were continued by John (Jack) Glover, who was responsible for the development and use of IIHR's new computer and who also developed new instruments and adapted them to feed data directly into the computer. Glover received grants for the expansion of computer facilities and for projects such as developing computer systems for acquiring experimental fluid-mechanics data. He also harkened to researchers who needed specialized instruments for their projects. In the ensuing years, two IIHR instruments were commercialized: Glover's Iowa Suspended Sediment Concentration Measuring System (an optical device for measuring temporal fluctuations of sediment concentration in turbulent flows, developed in the late 1960s) and Nakato's Iowa Rapid Sediment Analyzer (for measuring the size distribution, fall velocity, and specific gravity of moving sediment, developed in the mid-1980s). Grants continued to be received for designing novel equipment to meet IIHR's new research demands (such as the low-temperature facilities), and flumes and sensors continued to be designed and constructed for teaching and research at other institutions.

Push for Economic Development Products

Economic difficulties struck Iowa in the mid-1980s. These were a result of a precipitous agricultural economy that came to be known as the "Farm Crisis." Across the state, segments of the rural population were thrown into despair. The state's universities were beseeched to feed plans and products into Iowa-based industries in order to instigate economic recovery. This acknowledgment of the value of university-bred commercial enterprise came as a validation of Kennedy's efforts at IIHR, since he had been promoting the interplay of capitalism and scholarship for the previous two decades. IIHR had a few commercial projects ready to go.

The most promising involved structures to protect streambanks from erosion. Solutions to this problem had been sought for centuries. Long ago, the Chinese had armored Yellow River levees with woven willow containers filled with rocks. Later, British engineers had used similar structures—rock-filled wire enclosures—to guard against erosion of torrential, monsoon-glutted streams in India. During Rouse's tenure, the effectiveness of such "rock sausages" was modeled at IIHR, as was the formation of cross currents and scour at river bends in attempts to design structures to control river-bend erosion. And in the 1970s, IIHR students, using a curving flume, were examining the twisting secondary currents generated by accelerating water pushing outward along bends in a river, eating into its curving bank.

Thus, the laboratory was prepared when the COE came to IIHR in 1980 to request new techniques for halting severe erosion along California's Sacramento River, which in some mile-long reaches was annually washing nearly two acres of prime agricultural farmland into the ocean. Rather than armoring the river banks with rocks or other structures, Odgaard and Kennedy tackled the problem by investigating structures set *into* the water, devices that could offset the damaging secondary flows. Using models and computational studies, they discovered that a series of rigid rectangular panels or vanes placed in the river at a slight angle to the flow would counteract the river's natural torque by pushing the slower bottom water outward toward the eroding bank. This slower water would displace the rapid, twisting currents that caused the erosion and thus would halt the water's eating away of the river's outer banks. As an added advantage, these "Iowa vanes" were much more economical than most traditional anti-erosion devices.

By 1985, companies were inquiring about manufacturing rights

Starting in the early 1950s, IIHR researchers investigated the effectiveness of woven wire "rock sausages" in guarding stream banks from flood surges. Here a graduate student holds small-scale models of these protective structures.

for the vanes. Discussions were under way about patenting the vanes and using their manufacture as the basis for new industry in Iowa. The following year, a private corporation, Iowa Hydraulics Consultants, Inc. (later renamed River Engineering International, Inc.), was established to market the vanes through the local consulting firm Shive-Hattery Engineers. The firm might also market other IIHR devices such as the Iowa Rapid Sediment Analyzer and the helicoidal ramp dropshaft structure as they became available. Simi-

"Iowa vanes," practical erosion-control structures whose size and placement could be custom-fitted to a particular river bend, were developed and patented by IIHR in the 1980s and since then have proven their effectiveness in the U.S. and abroad. These vanes are being installed in Iowa's Wapsipinicon River.

lar commercial efforts were to be promoted by creating an endowed chair in hydraulics, to be filled by an established hydraulician with a bent for both academics and capitalistic ventures. This initiative proved difficult to accomplish and sustain, but it is noteworthy as an extreme display of IIHR's transformation since the Rouse years, when even whispered discussions about specific efforts and problems with industry were discouraged.

In 1987, the university obtained a patent for an improved vane: one with a double-curved surface that did not interfere with either the river's flow or its ability to carry sediment. Patents also were filed for the sediment analyzer and the drop structure. However, the dream of IIHR-spawned Iowa-based industry was not to be. River Engineering International survived for several years while marketing the vanes but then, before yielding a creditable profit, it folded.

Yet these projects were not considered failures. By 1985, IIHR, true to habit, had used the vane project to generate another type of research. Kennedy and Ettema had received a major grant to study, among other things, the efficacy of vanes in directing ice floes downstream and thus ameliorating destructive ice jams. Meanwhile devices modeled after the original Iowa vanes were—and still continue to be—used in numerous sites here and around the world to reduce scouring that threatens pipelines, roads and bridges, and farmland, and for other such river-training purposes. On the Nishnabotna River in western Iowa, among other places, vanes were found to work even better than predicted—reducing the river's usual annual inward migration through streambank erosion from 50 feet per year to only three or four feet.

Other Social Involvements and Historical Efforts

With IIHR's purposeful thrust into applied projects and real-world problems, it's not surprising that researchers were pulled into other matters of social relevance. In 1969, for example, five researchers were filmed for a public-relations movie on the role of navigation dams in Upper Mississippi River flooding. Later that same year, while protests against the Vietnam war were throwing universities across the country into upheaval, the university's student press proclaimed that an example of the "UI's contribution to the nation's war machine [could] be found in the Department of Hydraulics and Mechanics." The condemned research, funded by the Department of Defense under the Themis program, involved studies of nonlinear vibrations of military and other vehicles by the mathematician William F. (Bill) Ames (who had joined the staff in 1967 and was also looking at the presumably more innocuous propagation of waves in automobile tires). Although the university was called upon to terminate this research immediately, it continued on an expanded basis until 1975, when Ames left.

The following year, IIHR was pulled into yet another social problem: the arbitration between the State of Iowa and the Omaha Indians of Macy, Nebraska, who contended that a sizable section of the floodplain of the peregrinating Missouri River was in reality still their land. Kennedy and Jain were asked to testify about the nature of the river's wanderings in a lengthy and highly disputed courtroom battle.

Meanwhile, some researchers continued to assess the evolution of research and knowledge and thus kept alive IIHR's efforts in the history of hydraulics. Macagno's fascination with Leonardo da Vinci's efforts in flow and transport phenomena garnered him grants from the National Science Foundation and National Endowment for the Humanities in the mid-1980s. About that time, Kennedy also commenced research in the history of hydraulics that fed into continuing historical efforts.

Growth of Facilities

With the ongoing growth of programs, equipment, and staff, the perennial cries for added space were soon echoing through the halls of the institute. At first, space problems were appeased by the departure of the last of the federal agencies. The U.S. Geological Survey had located its offices in the Hydraulics Laboratory since 1932. In 1968, it vacated part of the third floor and all of the fifth, allowing room for new equipment such as a "snakelike channel" for water-pollution dispersion studies. In 1972, when energy-related studies were soaring in number, a building in Coralville was rented to house a large model of a power plant and its cooling pond, and the Hydraulics Laboratory received several large water heaters for modeling efforts of thermal effluents from such plants. The next year, the Coralville quarters were traded for rented space across the river, but people were more cramped than ever. "Our labs are now so short of office space that there is even a desk between the water tunnels of the first floor," complained the holiday letter, at the same time acknowledging that a "flume of giant size" was needed for modeling (within water) the flow of air around model cooling towers. Finally, in 1975, a new IIHR annex (today's East Annex) was completed across the river to house the large, 100,000-gallon Environmental Flow Facility, as well as cold labs for ice research.

Meanwhile the West Annex (earlier called the Model Annex) was failing rapidly. This annex, which had been constructed on a dump site across Riverside Drive from the Hydraulics Laboratory, was shifting and settling and had become too uneven for long flumes. As early as 1968, shortly after he came, Kennedy had started searching for funds for its replacement. It continued to be called into use, for example to house a thermal-pollution model of the Florida coastline, even while the building shifted, cracked, and a large mudslide crashed into its appended garage, taking out portions of the large wind tunnel. Still, the West Annex remained occupied because of lack of other space. Meanwhile, more new equipment was being constructed in the East Annex, for example another refrigerated room with a unique ice-towing tank and flume for studies of icebreaker ships (completed in 1979). That same year, the Oakdale 1 Annex for large hydraulic models was opened at the university's Oakdale Campus. It was immediately occupied by a model of a nuclear power plant and several miles of Korea's coastline that would be affected by its effluent.

Project number and diversity, equipment needs, and building space continued to feed one another, each pushing the others in an upward spiral. In 1980, Kennedy reported to the dean that without additional laboratory space, IIHR would need to start reducing its staff and phasing out various activities. Two more new buildings were forthcoming, both near the East Annex: the Model Annex in 1982 and the Wind Tunnel Annex in 1984, the latter holding the large low-turbulence wind tunnel (replacing the abandoned West Annex wind tunnel) and a new icing wind tunnel. The former, designed by Patel and partially funded by the National Science Foundation and the Office of Naval Research (ONR), enabled continuation of basic studies of turbulent flows, sponsored by ONR, while the latter was devoted to studies of near-freezing, moisture-laden winds and their ice loading on structures such as power lines and airplane wings—subjects on which little fundamental work had been performed. The West Annex was still employed, although it had settled so much that scientific instruments no longer could be properly aligned within its cracked walls, and staff had to be evacuated following large rainstorms because of landslide threats. Finally, in 1986, the building was relinquished to the university's Physical Plant to be used for storage, and in 1992 it was torn down to permit development of a green space. Thus, within nine years the institute had lost one laboratory annex but gained four large new ones, a sure sign of the financial as well as physical growth that had accompanied the tremendous surge in project number and diversity.

Staff members may always moan over problems with space. "We salivate over every nook and cranny that could possibly be used for office space," stated the Christmas letter in 1989, the year that the first-floor men's bathroom and shower were seized for library space. And in 1996, Oakdale 2 Annex was opened to house a salmon-study model, one of the largest hydraulic models ever constructed. This annex brought IIHR's total floor space up to 86,575 square feet, with the five laboratory annexes collectively providing twice as much floor space as the Hydraulics Laboratory building itself. The location of IIHR's buildings is shown on the map on the page following the preface.

Fundamental Fluid Mechanics and Ship Hydrodynamics

Fundamental research, which had dominated the Rouse years, continued but at a far lower level than before. Projects examining turbulent, stratified, and shear flows and their interactions and variations, never ceased. They were augmented by the 1971 arrival of V. C. Patel, an experimental fluid mechanician interested in applying three-dimensional boundary-layer theory to problems of viscous flow, and (with Macagno's retirement) a second experimentalist, Belakavadi ("BR") Ramaprian, in 1976, also interested in boundary layers and turbulence. Grants were received to pry into pulsatile flows, near wake flows, buoyant jets, and pressure distribution around rough-walled cylinders. In 1978, Allen Chwang, a theoretician, joined the staff. A strong applied mathematician, his interests lay in wave propagation and suppression, hydroacoustics, and hydrodynamics of highly viscous fluids. In stark contrast to a few decades earlier, when government funding for fundamental fluid mechanics could be obtained with relative ease, such funding had now become sparse. Ramaprian left in 1985 and Chwang in 1991. Two new fundamental fluid mechanicians have since joined the staff: Jeffrey Marshall in 1993, a specialist in vortex dynamics (see chapter 20), and Ching-Long Lin in 1997, who studies simulations of turbulent flows in the planetary boundary layer.

IIHR's fundamental studies have been closely linked to ship hydrodynamics, a research field whose evolution is traced in detail in chapter 19. Ship-related efforts continued to flourish and receive regular funding from the Navy. Louis Landweber's leadership in the field centered on his work with wave and viscous resistance and his pioneering efforts to develop ideal-flow theory for application to ship hydrodynamics. Another pioneering effort commenced in the 1970s, when Patel started numerical modeling of viscous flow around ship hulls, supported by experiments in the wind tunnel. After Landweber's retirement, IIHR received major funding to develop further both computational and physical models relating to flow around ships, efforts that have led to IIHR's establishment, under Fred Stern, as a leader in investigations of viscous ship hydro-dynamics and free-surface flows. Research has utilized both experimental and numerical modeling techniques, the latter constituting the discipline of computational fluid dynamics (CFD). CFD techniques, which were nurtured at IIHR primarily through ship hydrodynamics, have since spread to many other IIHR initiatives and have become a major and unifying thrust of the 1990s. The development of CFD and its links to ship hydrodynamics are outlined in the following paragraphs.

The Impact of Computers; Development of Computational Hydraulics (CH) and Computational Fluid Dynamics (CFD)

The many research foci described in this chapter have produced a fluid and ever changing mixture that continues to evolve in response to societal needs and funding availability. This evolution has perhaps been shaped by a single instrument—the computer—more than anything else. Increasingly powerful and sophisticated instruments interlinked with computers are now tied to every phase of the research process, from acquisition and recording of data to analysis and presentation of written and visual research results and their communication among colleagues.

IIHR's first computer and Jack Kennedy arrived at IIHR nearly simultaneously. Glover had worked for years to acquire that first in-house computer, which arrived in 1967 and was funded in part by the U.S. Navy primarily to allow the recording and analysis of turbulence data. Within months, IIHR researchers realized that the computer would significantly impact other areas of research, and a year later Kennedy reported that the "favorable impact this facility is having on practically every phase of our research program exceeds even our most optimistic expectations." Not only were data acquired and recorded far more rapidly than anyone could have imagined; the computer's processing capabilities were "leading to new interpretations of results" that were previously unobtainable with analog techniques. New types of projects were generated as well. For example, in the late 1960s, IIHR researchers undertook the comprehensive plotting of sedimentary dunes and ripples in the Missouri River's bed between Sioux City and Council Bluffs, Iowa, a highly publicized project funded by the COE that attempted to re-

John Glover (standing) and Subhash Jain (on carriage) participated in one of IIHR's very first computer-dependent efforts, which involved mapping the flow distribution and alluvial-bed conditions in a reach of the Missouri River. The formation and effects of ripples and sandbars under varying flow conditions were examined in a 120-foot-long tilting meander flume. A sonic sounder rode the motor-driven carriage above the flume, linked (through a converter) to the IBM 1801 computer (inset), which processed the data. The volume of data collected would have made this study impractical prior to the computer age.

late the riverbed's shape to its flow, resistance, and the timing of its flooding. This was followed by modeling the formation of ripples and dunes and studying their relationship to water flows at IIHR. Before the computer was available to record and process the large amount of necessary data, this project would have been unimaginable.

The computer quickly became an object of research in itself, with Glover applying for grants to develop programs and to upgrade and expand the computer system. The institute's holiday letters reported that these expansions were causing Glover to act like a kid on Christmas morning. Soon he was reporting that research areas that had previously seemed hopeless were now opening up. While the computer's ever-greater integration with the research process was obvious, perhaps no one suspected how the computer's increasing power, speed, and graphical capabilities would turn IIHR's research in a totally new direction, that of numerical modeling. That realization became evident through two parallel fields of exploration, the CH efforts led by Forrest Holly, who arrived at IIHR in 1982, and the CFD efforts instigated by V. C. Patel.

Holly, a civil engineer, came here from the firm SOGREAH Consulting Engineers in Grenoble, France, one of the world's first organizations to attack engineering problems with CH: the development of numerical techniques for producing computer-based simulations of fluid flows and for solving hydraulics problems. He arrived just a year after Kennedy and his student F. Karim had begun numerical simulation of bed degradation in the Missouri River, a study that became a precursor to what is now called CH. Holly helped introduce CH to the U.S. He brought to IIHR a leadership role in a new focus area as well as licensing agreements to use SOGREAH's computer software for educational and engineering activities.

Holly has consistently emphasized the development of computer software for applied engineering practice. Through Holly's programs, IIHR's hydraulics consultation efforts have taken on a new meaning. In addition to visiting a site to devise solutions to a given problem, Holly might provide industrial sponsors with a computer program that on-site engineers would themselves apply to their particular situation—for example, to input location and water flow of a certain irrigation canal, push a button, and receive information on the effects of the surge of water that will be created if a nearby water gate is opened. Holly's programs attack a number of practical problems: flood-wave propagation in rivers, surges in irrigation canals, storm-drain design and operation, dispersion of pollutants in rivers, sediment transport, and thermal responses of rivers, to name some. Holly's efforts in the 1990s have been focused primarily upon perfecting CanalCAD, a program to simulate the dynamic flow of water in irrigation canals and to improve on-site management and minimize water wastage of irrigation systems, and CHARIMA, a computer code that Holly calls his workhorse. The latter general-purpose code simulates the movement of steady or unsteady water, sediment, and contaminant movement in simple and complex river channels. Holly continues to use CHARIMA to monitor the dispersion of heated wastewater in the Upper Illinois Waterway (as well as elsewhere) for Commonwealth Edison, the company that initiated IIHR's large-scale efforts in power-plant studies back in 1970.

Several features distinguish IIHR's work in CH. The field originated among civil engineers and their dealings with free-surface flows, in particular those of rivers. Because CH simulations address natural flows and the inconsistencies of nature, and because CH models attempt to simulate changes over long time periods—months to years—over large areas, the models have been limited to a single dimension, viewing a river, for example, as a thread without thickness. In the 1990s, CH has been expanding into two-dimensional models (that regard a river as a ribbon of varying width) and three-dimensional models (that see the river as a tube with both width and depth). Holly's CH programs are regularly sold through IIHR to industrial users. The orientation toward practical applications has led Holly, whenever possible, to adapt his codes for small desktop PCs that are readily available to practicing engineers (although his codes typically have been developed on larger, more powerful computers).

By 1985, with the growth in computer speed and power as well as growing familiarity with computer capabilities, several IIHR staff members had been pulled into CH operations: Kennedy, Nakato, Jain, Odgaard, Schnoor, and Georgakakos, as well as Holly. But by that time a sister science, CFD, had also become a major component of IIHR research. CFD gives an extremely detailed view of what CH paints in broad brushstrokes. As the mechanical engineer's correlate of the civil engineering–dominated CH, CFD was born in the aerospace industry and, driven by critical needs and strong financial support from the defense program, made major impacts on nuclear and mechanical engineering. CFD has its roots in fundamental studies, looking at the fluid flow in and around simplified shapes and exploring the intricacies of turbulence. When applied to real-world problems, CFD traditionally has avoided natural flows and dealt instead with industrial flows and flows around or through human-made bodies: a ship's hull, an airplane foil, a turbine. Its examination is limited to a small area and to conditions that do not change over time. With its attempts at extremely high resolution of fluid movement, CFD is always at the forefront of numerical modeling. It requires the most powerful and sophisticated supercomputers of the day and constantly demands more of this technology. However, even when using state-of-the-art machines, CFD is limited in its capabilities by computer power and speed. CH, in contrast, is limited by the inherent "messiness" and ever-changing complexity of natural situations.

The roots of IIHR's CFD efforts were planted in the 1960s, about a decade before the term CFD was coined, and are embedded in the pioneering work of Enzo Macagno and his graduate students. They used the university's mainframe computer to solve the Navier-Stokes equations —equations of motion for incompressible fluids— applied, for example, to fluids rapidly expanding in pipes and rotating in cylinders. About the same time, Landweber, with the computational help of Matilde Macagno, was challenging the computer to find numerical solutions to ship wave resistance problems using theories of ideal fluids. These early forays were precursors to the systematic efforts of Patel, who established CFD as a major IIHR research field and tool.

Patel had co-authored a book with John Nash entitled *Three-Dimensional Turbulent Boundary Layers*, published in 1972. That book outlined a vision for merging ideal-flow theory, typically attacked with elegant mathematical formulae, with boundary-layer theory, through which the viscous effects of real fluids could be interpreted. Until that time, real flows were studied only experimentally or in highly simplified geometries because of the impossible complexity of their numerical solutions. Patel and Nash's book, which outlined techniques for their numerical solution, became a significant contribution to the field. It also outlined the methods that Patel started to apply at Iowa to the boundary layers of ships (ignoring the effects of surface waves), toting his boxes of computer cards to the university's Weeg Computing Center, plotting the outputs on graphs by hand, and comparing his results to experimental studies completed in IIHR's now-defunct large wind tunnel in the old West Annex. These investigations into numerical interpretations of complex, three-dimensional flows were less than ideal. The early computers could not produce the graphical aids necessary for Patel to really comprehend his results. Also, the simplified versions of the

Navier-Stokes equations demonstrated very severe limitations when applied to flows over ships, airplanes, and other actual objects.

Spurred on by these problems and by the computer's growing capabilities, Patel expanded his techniques and, in the early 1980s, became one of the nation's first researchers to use the computer to apply the full-blown Navier-Stokes equations, coupled with turbulence modeling, to flows around ships and similar three-dimensional shapes. The application immediately opened up all sorts of possibilities in terms of studying viscous flow around the full ship, including its propeller and wake. These efforts received a significant boost when Patel attended an International Towing Tank Conference in Leningrad in 1981. There, U.S. Navy personnel were fretting about certain capabilities of Soviet submarines that Patel attributed to the Soviets' advanced flow-prediction technologies. When asked what was required to duplicate Soviet abilities, Patel conceived a comprehensive research program to model the flow of water around a ship, from its bow to its wake. The result was a major long-term Special Focus grant from ONR, received in 1982, to conduct basic research in computational and experimental modeling of fluid flows around ships, including consideration of factors such as the friction on the hull, wave generation at the bow and stern, and turbulence in the wake.

This grant facilitated the full-scale establishment of CFD at IIHR. The grant led to the production of RANSTERN, one of the first successful computer codes for calculating viscous, three-dimensional turbulent flows around ship sterns and wakes. RANSTERN was developed by Patel and his post-doctoral associate Hamn-Ching Chen—a former student of IIHR researcher Ching-Jen Chen, who was instrumental in developing some of the initial algorithms used in IIHR's CFD models. RANSTERN became a prevailing CFD methodology in the 1980s and since then has fed into other related computational efforts. Perhaps more importantly, the RANSTERN code played a key role in opening the minds of U.S. Navy hydrodynamicists to the notion and usefulness of CFD modeling. In doing so, it became a milestone in shaping the evolution and funding of ship-hydrodynamics research.

Patel has subsequently devoted himself to other turbulent-flow problems. With the help of post-doctoral associate Fotis

CFD models today are capable of tracing complex flows around three-dimensional bodies, resulting, for example, in these interlocking hairpin vortices shed from a sphere moving through a fluid. These modeled vortices are similar to those produced by air flowing around a fast-moving golf ball or baseball.

Sotiropoulos, he started applying these techniques to a variety of other complex fluid-flow quandaries, producing numerical simulations of three-dimensional flows through the human voice box, through hydropower installations, and around simple, fundamental bodies. Their efforts were enhanced by links to the nation's supercomputer centers and by the ever increasing power of computers, as well as their expanding ability to depict their results in detailed graphics.

Meanwhile, Fred Stern, a naval architect who had joined the research staff in 1983, assumed leadership of IIHR's ship-hydrodynamics efforts and started to develop CFD codes for free-surface flows (that is, to include consideration of waves generated at the boundary between water and air), as well as for more detailed deliberations of propulsar-blade flows (a topic described in greater depth in chapter 19). Since 1994, IIHR's ship-hydrodynamics efforts have taken on the additional task of developing general-purpose CFD codes that would be useful in research and development in academia, industry, the U.S. Navy, and elsewhere. The efforts of Stern and his research team have continued to energize IIHR's CFD research, initiating connections with Department of Defense (DoD) supercomputers, acting as training grounds for students, bringing in major funding in the field, and always demanding the maximum power that computer resources are capable of supplying. In 1996, this research team was chosen to conduct one of a select number of DoD Challenge Projects, a designation that provides extensive access to some of the most powerful supercomputers in the world. The award inaugurates researchers here as those addressing DoD's most pressing technology challenges, and as leaders who are developing computational methods that will become an integral part of virtually every defense mission area. Through projects such as this, ship hydrodynamics remains at the foundation of IIHR's CFD investigations.

1991 to 1997

In 1991, with John Kennedy's death from cancer, IIHR again lost a strong-willed and directive head, just as it had nearly 60 years earlier when Floyd Nagler had died of a ruptured appendix. However, as was true at Nagler's death, firmly established research directions held the institute on a steady course. Rouse's rigid adherence to fundamental fluid mechanics now had mellowed into a rich, diverse, and prolific mixture of applied and fundamental studies. During the quarter century that Kennedy had directed IIHR, the number of research grants had grown from 21 (in 1967) to 76 (in 1991). The 1991 budget totaled around $2,930,000, nearly a 12-fold increase from the approximately $250,000 annual budget at Kennedy's arrival. In stark contrast to that earlier year, when all research was supported by only 12 funding sources, 10 of which were governmental, and half of all funding came from the U.S. Navy, research funding now was received from over 40 public and private granting and contract agencies. This diversity of funding sources, project types, and applied and basic research had strengthened IIHR's resiliency, enhancing its ability both to withstand the vagaries of research funding and to walk at the forefront of new research initiatives.

The 1991 projects reflected the several new fields that Kennedy and colleagues had coaxed into existence here, with many but not all projects still relating to the environmental theme that had blossomed soon after his arrival. Of the 76 projects in 1991, 17 related to power plants in some way. A number of these incorporated hydraulic models. Many were concerned with questions of thermal diffusion or sediment problems. Ship hydrodynamics claimed about a dozen grants, many of these dealing with questions of fundamental fluid mechanics and CFD, each of which was also represented by a few additional grants. Hydrometeorology projects numbered nine. Several fields were represented by four to six projects: ice research, computational hydraulics, salmon and sediment studies. The salmon research, ship hydrodynamics, and hydrometeorology brought in a major amount of IIHR's funding then and remain a financial mainstay to this day. Additional projects covered a diversity of fields, among them educational initiatives, history of hydraulics, development of equipment, dropshafts and other hydraulic-model work, and water-quality investigations.

By 1997, the annual budget had increased another million dollars, to $3,951,000, with the number of projects fluctuating between 74 and 91 in the 1990s. The details of each research field continued to evolve, but general subject areas remained remarkably constant. The few fluctuations reflected alterations in funding sources and personnel. The number of hydrometeorology projects nearly doubled. Ship-hydrodynamics projects dropped almost 50% in number but maintained approximately the same total funding. While many projects centered around hydraulic models of various structures, the number of all projects associated with power plants plummeted from 17 to 7. Basic-fluid-mechanics efforts encompassed studies of vortex dynamics.

IIHR in 1997 represents one of the few sites in the world where mechanical and civil engineers work together on a daily basis, and where both CH and CFD have been highly developed. The infusion of these techniques into every aspect of research may become the hallmark of IIHR's research in the 1990s. With it has come the breakdown of boundaries between CH and CFD and the merger of these previously disjunct areas of endeavor. CFD is becoming more applied in nature and is starting to address the complexities of the natural world, being pushed outward in space and time by the multiple hydraulic-engineering problems that are posed through CH. CH, in turn, is profiting by the increased computer power of smaller computers and, stimulated by the sophisticated techniques developed through CFD, is tending toward more detailed, three-dimensional modeling. In a sense, this fusion of applied and theoretical approaches is analogous to that which occurred when Hunter Rouse started to integrate fluid mechanics and hydraulics back in the 1940s. Those two fields also begged for discussion between mechanical and civil engineering; neither was progressing to its fullest extent without the other. Such dialogue is creating tremendous opportunities for a critical merger of numerical-modeling ideas and skills, potentially leading to the production of accurate models capable of predicting the details of hydraulic events in complex natural settings.

That such a merger is occurring as the twentieth century closes is displayed by several computer-simulation projects. Applying these CH-CFD techniques, IIHR researchers are looking at detailed, three-dimensional flow processes over dunes and are examining pollutant dispersion and mixing processes underneath ice. Power-plant facilities are being modeled in detail: researchers are piecing together programs that will represent the detailed modeling of water flow through a river and reservoir, into hydropower intakes and through turbines, and back out and down the turbulent, highly variable river. Such three-dimensional modeling, which poses a formidable challenge to researchers, is being done for the Wanapum Dam on the Columbia River, a site of intensive salmon studies. It promises to supplement IIHR's extensive work with physical models of this site and also to be nurtured by physical models, as a new model of the Wanapum Dam site has been constructed to validate IIHR's numerical-modeling efforts (see chapter 15). This numerical model is especially significant because it is thought to be the first comprehensive attempt to numerically simulate the complexities of natural river geometries. Numerical simulations are also being applied to studies of power-plant intake structures, sediment transport and scour, and fundamental studies of vorticity. Hydrometeorological research also relies on computers to interpret complex fluid-movement problems through quantifying uncertainty and random processes connected with weather forecasting.

With computer applications consuming the creative energies of many IIHR researchers, one might wonder if hydraulics laboratories in the future will be dominated by computer programmers who never really get their hands wet. Are physical models and laboratory experiments destined to become anachronisms?

It is true that numerical models and their manipulation, viewed on the computer screen, promise to replace physical models in some situations and to be employed for hydraulic problem-solving and structure design. Several advantages warrant a move in this direction: once reliable numerical models have been developed, they are likely to be easier, faster, and cheaper than their physical counterparts. They also can overcome limitations of scale that sometimes hamper hydraulic models. Sand particles, for example, can be scaled down only so far. However, while the field of numerical modeling is developing extremely rapidly, it will still require considerable vision and effort to reach the point of everyday employment, and industrial contractors (who are sometimes skeptical about this technique) will need to be educated about the potential of numerical modeling. IIHR can provide professional leadership on both counts. Even after these barriers are overcome, hydraulic models will remain necessary for validating and calibrating numerical models. And with the best of intentions, some problems that remain too complex and variable for numerical modeling—such as the details of real-life pump intakes—will continue to require observation and manipulation within physical models.

Experimentation, hydraulic modeling, and more intimate hands-on manipulation are bound to remain crucial for another reason: because they guide the human brain to see solutions not perceived on a two-dimensional computer screen. Kennedy foresaw the threat of losing a physical intimacy with the river and its measurement when the first computer was installed at IIHR back in 1967. At that time, he simultaneously lauded the IBM 1801 computer and warned of an attendant hazard, the "loss of the physical insight and germs of ideas that are often born of a thorough familiarity with a set of data, a familiarity that can only be obtained by working intimately with it." Twenty years later, he applauded a productive synergism that he saw increasingly dominating the hydraulician's routine, in which experimental fluid mechanics played a vital role "supplying the necessary inputs for numerical models, investigating phenomena that are too complex to be attacked numerically, and guiding the development of numerical models." He envisioned a future in which "physical and computational models will be integrated into complex units, with the computer both analyzing the output of and directing the course of experiments." Kennedy recognized that the richest productivity results from interplay between physical and numerical models, with researchers going back and forth—observing a process, trying to model it numerically, then returning to the real-life setting to determine whether what was modeled is actually happening.

Thus, numerical modeling is likely to remain one of several tools in the hydraulician's toolbox, supplemented by physical models and field observations. Just as in past decades, IIHR's most significant future contributions are likely to result from the continued synthesis of the theoretical and applied, the continued mixing of researchers with a diversity of training, skills, and outlook, and the continued and simultaneous use of numerical and physical models.

14. Moving into the Computer Era

"I'd run across Riverside Drive to the West Annex, get all my sediment analyzers set up in the oscillating flow flume, then run back across the street to the Hydraulics Lab and flip the switch to turn on the computer. Back across Riverside to do a run. Across again to the lab, clear the disk by spilling the data onto punch cards, turn the computer on again for the next run. Back across to the West Annex to do another run. Back to the lab . . . ." Thus IIHR research scientist Tatsuaki Nakato explains the contortions he had to go through back in 1973, when he was using IIHR's first computer to gather data for his dissertation on sediment entrainment. On the precious few days when everything worked well, he could gather useable data almost continuously. But more often, something went wrong. The shop crew might be using a welding torch or other power tools to construct a model, for example, and a voltage surge would cause incorrect data to be fed into the computer. A bolt of lightning might shut the computer down entirely, leading prudent operators to simply turn the computer off when storms threatened. At other times, the computer would shut itself down and refuse to record data even though everything seemed to be in working order. "The connection between the computer and the experiments was very flaky," explains Nakato when describing the data-collection system, which relied upon a large data cable that ran between the Hydraulics Laboratory and the West Annex through a conduit underneath Riverside Drive. "The data-collection system was temperamental. When something went wrong and the system crashed, getting it back up and running again could be a nightmare."

That was the IBM, more formally called the IBM 1801 Data

IIHR researcher Art Giaquinta is using sense switches to program IIHR's first computer, the bulky IBM 1801 acquired in 1967. The complete IBM was three times as large as the portion shown here, yet only a single user could access it at a time.

Acquisition and Control System. Laboriously slow in its operations, the colossal instrument possessed but a tiny fraction of the memory and disk space of laptop computers that today's researchers tuck like notebooks under their arms. Yet the procurement of the IBM 1801 in 1967, the year following John Kennedy's arrival as director, heralded a major thrust forward in the institute's capabilities. Prior to that, researchers had been able to use only the university's centralized IBM computer at Weeg Computing Center for calculations. But with the IBM's arrival, graduate students started to celebrate their liberation from the painstaking tedium of reading and recording multiple measurements by hand—when the computer system was working properly, that is, and when you had designated time on the computer. The latter was not to be assumed. Only one person could use the computer at a time, and professors had first shot at the daytime slots, which meant that graduate students using the computer often worked through the night. They considered themselves lucky to do so, for Jack Glover, who was in charge of the IBM and its scheduling, was a strict manager and meticulous caretaker whose lengthy list of "dos" and "don'ts" had to be followed if the computer room was to remain unlocked and accessible. If someone was to violate his operations code, bring food or a bottle of pop into the computer room, for example, that student might suddenly find it very difficult to get scheduled computer time. Sitting for hours typing complex programming sequences onto IBM punch cards became a privileged act during those early days of computer use. Students earned the privilege by giving the computer and card puncher a weekly washing with window cleaner, rags, and Q-tips. "Glover really taught us discipline," remarks Nakato.

In 1978, Glover took a year-long sabbatical at the Hewlett-Packard main office in Palo Alto, California. During his absence, the institute received an NSF grant to purchase a new computer, and upon returning Glover oversaw the 1979 acquisition and installation of an HP 1000 E-series minicomputer. The HP immediately altered the ambiance of the laboratory. For one thing, researchers who had gained a reputation as "computer hogs" by pirating the time slots of other users were offered a reprieve: the HP was a multi-user system, meaning that it could be accessed by multiple researchers and projects simultaneously. Not only did it look very different from the IBM—as a "mini-computer," it was perhaps a quarter of the IBM's size—it also had capabilities never dreamt of with the IBM, such as the ability to plot data on a graph and to control portions of an experiment. But it had some shortcomings. As the core of a centralized computer system, it still was connected by underground cable to the Model Annex and to other sites with experimental equipment. The cable connections still frequently gave problems. And like the IBM, the HP depended on a single person who knew its idiosyncrasies and kept it running. As electronics engineer Doug Houser put it, "If the computer broke down, all data acquisition stopped and everyone went on vacation." He recalls times when the HP would flounder and the systems manager, Jim Cramer, was out of town. Houser and Cramer would then spend hours on the phone, Cramer troubleshooting from a distance by suggesting operations that Houser entered into the computer, reporting back to Cramer on their results.

Nevertheless, the HP was faster and more powerful than the IBM by a wide margin. Jacob Odgaard recalls how two of his own projects exemplified that change. In 1980, just a few years before the IBM's demise, Odgaard used the IBM in a model study of thermal plumes created by heated effluents from power plants. While Odgaard sat next to the computer, shop manager Jim Goss was stationed next to the model at the Oakdale Annex, six miles to the north. Goss would manually position more than 100 thermistors, instruments that recorded water temperature. When done, he'd call Odgaard who would turn on the computer, take a set of readings, and examine the printout of raw data. If the data were satisfactory, Odgaard would call Goss back and instruct him on where the thermistors should be relocated so the next run could commence. Back and forth the two would go, completing round after round of measurements, usually between 10:00 p.m. and 10:00 a.m. in order to avoid the annex's intense heat. In stark contrast, about five years later, Odgaard was using the HP to complete the institute's first totally computerized model study, one examining sediment movement in a curved channel. A motorized carriage bearing instrumentation moved over the Oakdale Annex model, automatically extending a probe that measured water depth first in one location, then in another. For each site and depth, a current gauge would record the water's velocity at each of ten equally distanced positions in the water column above the sediment. The siting of the carriage and instruments, as well as the recording of data, were all controlled through a complex program run by the HP, which remained at the Hydraulics Laboratory but (unlike the IBM) could be operated through a remote terminal and modem (telephone) connection.

Changes in computer capability such as this presaged a never-ceasing revolution of technologies and personnel. Within a few years of the HP's arrival, the IBM's mammoth steel carcass had been hauled away, making way for a sequence of ever smaller, faster, more user-friendly, and more powerful computer systems, each compounding the capabilities of its predecessors.

To use the IBM, a researcher had been forced to sit next to the behemoth and toggle its switches. The HP could be accessed far more easily, through satellite terminals, so that a researcher could

Unlike the IBM, the HP could be accessed through satellite terminals via modem connections. IIHR purchased a second HP in the mid-1980s for running the laser-Doppler velocimeter installed in the newly constructed low-turbulence wind tunnel (as well as other equipment in the Wind Tunnel, East, and Model Annexes). The terminal shown here controlled the off-site HP computer, which received data from transducers in the wind tunnel (after the raw data signals were amplified and processed by equipment on the electronics rack to the left).

tap into the HP from any office or laboratory with a terminal and modem connection. By the mid-1980s, large multi-user computers in centralized computing facilities, with their frustrating connections to satellite terminals and laboratory equipment, were being replaced by computers that had shrunk to fit on the researcher's desk. These versatile new computers were plummeting in size and cost even as they expanded in computing power, ease of use, and speed—to the point that within a few years, a single desktop computer could rival the centralized HP facility for most functions excluding the sharing of acquired data among different users. Yet they remained dedicated to the individual researcher or project and could be programmed to fit that individual's needs.

These rapid developments encouraged Robert Hering, dean of the College of Engineering, to take a bold step to assure that the college's students and faculty possessed state-of-the-art computer equipment. Realizing that computer availability and literacy were becoming an assumed requisite of a quality engineering education, Hering raised funds in the mid-1980s to purchase computers for college faculty members. His choice was the Macintosh (Mac), which had first rolled off the assembly line in January 1984. Some Mac computers also were purchased for placement in newly created student computer laboratories in the Engineering College and an existing computer lab at IIHR, under the aegis of ICAEN, the Iowa Computer Aided Engineering Network, formed in 1985 for the promotion of undergraduate education.

Throughout the college, most of the computers were carried into faculty members' offices, where they found mixed acceptance. Some experienced computer users zealously integrated the Mac's word-processing and drawing abilities into their preparation of teaching materials. Some less experienced users tried the computers, decided to stick with pencil and paper, and presented the machines to their secretaries. A few let them sit in their boxes untouched until they were picked up several years later, when Dean Hering presented faculty with new computers. Whatever the response, the handwriting was on the wall: small office computers were the next wave of the future.

In 1986, Nakato and Forrest Holly became the first at IIHR to operate an IBM personal computer or PC (which was similar to the Mac in function but had a different operating system and software). PCs and Macs have remained the institute's most abundant workhorses ever since. Today they are found in nearly every IIHR office and laboratory, accompany researchers into the field, and are carried along on travels as laptops. They serve as word processors, data recorders, and computational instruments, and as Internet electronic linkages to information and colleagues around the world.

By the end of the 1980s, IIHR's in-office computers were joined by another instrument far more powerful (and also more expensive) than PCs and Macs, the UNIX workstation. The greater speed and memory of workstations allow them to perform much more complex computations and express the results as sophisticated graphics. Workstations remain dedicated to a particular researcher but can be linked to other computers and equipment electronically, and thus they can share common resources: for example, several units could tap into one another to share computing power, thus multiplying the processing power of any one unit.

Even the workstations were not powerful enough for researchers involved with computational fluid dynamics (CFD), who were creating sophisticated three-dimensional numerical models for fluid motion. From the mid-1980s onward, these researchers started to use yet another type of computer, the extremely powerful but equally expensive supercomputers that were then becoming available to universities through distantly located supercomputing centers run by the National Science Foundation or Department of Defense. IIHR researchers with the most challenging computing needs soon became the university's major users of these centers and remain so to this day: in the academic year 1994–95, for example, IIHR researchers accounted for 92% of the university's allocation for supercomputers.

While some IIHR researchers ran programs too large to be computed by anything other than a supercomputer, use of government centers was not always convenient or timely. To keep research schedules running well and to augment use of the supercomputer centers, IIHR decided to purchase its own supercomputer. In 1996, a Silicon Graphics Power Challenge Array was installed, its sixteen 75 and 90 MHz MIPS R8000 processors sharing 4 GB of RAM and 60 GB of disk space. This supercomputer placed IIHR in the top tier of university computing centers nationwide. It made computer usage once again more local and thus more convenient. With the new computer, IIHR's in-house computing power, which had been increasing rapidly through the 1990s, was multiplied five-fold in 1996. The $2.1 million system was purchased with funds from the university's central administration and IIHR supplemented by a grant from Silicon Graphics.

The new supercomputer is accessed via workstations, PCs, or Macs through networking—that is, using a local computer to tie into resources around the world (as well as other local computers). The trend was started by the Department of Defense, which in the early 1980s first allowed citizens to use a prototype of the Internet to connect universities around the country to its supercomputer centers. The multiple advantages of networking soon became evident. Not only could general information and publications be located and shared with ease among colleagues and institutions, professionals could instantly interact with one another through electronic mail or e-mail, which can transfer entire data sets and large written documents as well as personal messages. The further development of networking systems forms much of the leading edge of computer technology. This and related technological developments continue to present novel applications for computers, making definition of the status quo ever more elusive and ephemeral.

This ongoing revolution of computer systems has been accompanied by a rapid evolution in IIHR staff. Glover brought computers to IIHR in the 1960s and then tended the IBM for over a decade. His acceptance of a permanent position with Hewlett Packard in August 1980 left IIHR with a new HP 1000 but no one able to operate it. Nakato agreed to pitch in, attending a few month-long training sessions at the HP office (returning from one to find his first daughter had just been born) and adding systems manager to his already lengthy list of job duties. "Students and faculty were shy about using the new computer. I had to install a few games to tempt them into trying to use it," he admits.

Nakato willingly relinquished these tasks in June 1981, when for the first time the institute hired a professional engineer specifically to oversee its electronics equipment. Jim Cramer, trained in electrical and computer engineering, became manager of computer systems and electronics laboratory instruments. Cramer dedicated most of his time to computers, receiving help with electronics instruments first from a student assistant, then from staff member Ken Hartman, and then from electronic engineer Doug Houser.

Meanwhile, Cramer's time was increasingly utilized by the hydrometeorology group, whose work consistently required far more computer capability than the institute possessed. In 1989, the group sought and obtained NSF funds to purchase UNIX workstations dedicated to hydrometeorological research. That same year, the group installed its equipment in a separate, newly established hydrometeorology laboratory in the Engineering Building. Cramer designed and brought up the laboratory and then became its systems manager. Later that year, Mark Wilson was hired as the new IIHR data-systems coordinator, and to this day he manages IIHR's general-computing laboratory and the many other computers sprinkled throughout the Hydraulics Laboratory and its annexes. In 1993, the Computational Laboratory for Hydrometeorology and Water Resources (usually called HML) was moved to the Hydraulics Laboratory, where it remains as a distinct entity within IIHR's computer system. It was managed by Anton Kruger between 1992 and 1996 and since then has been managed by Wilson and his support staff of one assistant and several computer-science students, with Kruger and others in the hydrometeorology group providing direction and oversight.

Today, IIHR's computers perform everything from word processing to information retrieval, communications, data collection, and highly technical computations. In many offices multiple computers are becoming the norm, each with its specialized function. PCs or Macs often sit next to workstations.

Their peaceful coexistence and abundance indicate the degree to which computers have infiltrated every aspect of IIHR's efforts and operations. As long as one has access to a computer terminal or carries along a laptop, communications remain open, and often nearly instantaneous, around the world. Floppy disks replace briefcases heavy with papers. Disks also simplify the export of computational hydraulics codes to industrial clients. Through e-mail and the Internet, our connectivity—with information, each other, and our work—has multiplied rapidly.

The result is sometimes a mixed blessing. Researchers now can perform and monitor their experiments in comfort from afar. The possibilities for collaborative research and long-distance teaching appear to be limitless. Professors are tied via e-mail to colleagues and students and administrators to their problems, no matter what the time of day or how distant they may be from the site or person in question. The accelerated pace of receiving information transmits to an accelerated rate of responding. This can lead to increased productivity. It also can reduce the thought given to an idea before it is shared with others. Some find the borders between their professional and personal lives murkier than ever.

No one would dispute that many menial tasks have been eased tremendously because of the computer and related technologies. However, while these sometimes increase efficient production or reduce stress, they also intensify the sense of urgency regarding grant applications and other dated materials, with materials that used to be rushed to the post office five days before a deadline now being submitted on the deadline through electronic mail. The rush sometimes leads to nostalgia, such as that expressed in IIHR's 1991 Christmas letter, which stated: "Sometimes we can't resist longing for the days of slide rules, pitot tubes, and graduate students staying up all night entering experimental results on data sheets. Life was so simple in comparison when staff meetings were called to announce the loan policy for the lab's three programmable calculators."

On a more mundane level, computers have transformed the institute's day-to-day logistics by replacing the typewriters and drafting tables of two decades ago. Mike Kundert, who executes the institute's drafting and graphics work, now does so on a computer screen. And Darian DeJong, an engineer who serves as a liaison between shop personnel and researchers, uses a computer to design equipment and to trace out templates for models that the shop will construct.

With their ability to manipulate words and equations, computers have eased the writing process and report preparation tremendously, both among researchers and in IIHR's main office. This transformation commenced with the arrival in 1982 of IIHR's first word-processing system, the NBI 3000 that came complete with 8-inch floppy disks. These systems were replaced in January 1989 by Macs purchased for the office staff, who were forced in one weekend to transform themselves into computer-literate users. Meanwhile, professional staff started to use their own computers to prepare correspondence and reports.

Word processing was predicted to reduce the need for office support staff, cut paper use, and instill similar economies. Secretary Twila Meder comments that computers and related technologies seem to have transformed habits rather than reduced needs. "Faculty who used to hand-write their letters and reports and give them to us to type now type them themselves," she concedes, "but our work load has not decreased, it's only changed to different

IIHR's first word-processing system was the large NBI whose bulky printer emitted a penetrating clatter. Office staff (such as Karen Nall, shown here) were thrilled with its 1982 arrival because the NBI could integrate equations into the text and print a report in a single pass.

activities." Nevertheless, with computers, IIHR's office staff has not needed to increase in size, despite the growth in research staff and publications. In addition, graduate students are now able to produce their own dissertations on the computer and thus are saved the cost of hiring a typist.

Teaching practices have benefited from the ease of preparing and transmitting documents, which has greatly facilitated preparation of classroom material and tests. An example of future trends may be the Computers in Civil and Environmental Engineering class taught by Witold Krajewski in 1996, which gained renown as a "paperless class." Wanting to force his students to become more relaxed and literate in their use of computers, Krajewski placed all course and lecture materials on the Internet's World Wide Web and assigned and collected all homework via e-mail.

Computers and related technologies have also become subject

matter in themselves. Teaching about the use of various technologies is nothing new. Back in the mid-1930s, for example, Chesley Posey wrote papers on classroom instruction in use of the slide rule and using slide rules for routing floods through reservoirs. Forty years later, Glover and Tom Croley were playing a similar role. In 1971, Glover was teaching workshops on computer-aided data acquisition for fluid mechanics, a field in which IIHR had become widely recognized. Croley, a water resources systems analyst who had joined the staff in 1972, became expert in programing the small calculators that were then in vogue. These instruments were extremely significant to engineers. They replaced slide rules and provided a relatively inexpensive tool for performing rapid and precise calculations whenever and wherever one pleased. Croley developed hundreds of hydraulics and hydrology programs for programmable calculators and wrote two widely accepted and much-used books filled with his programs. He also taught several short courses on how to use programmable calculators and his programs. His books were published in 1977 and 1980 by IIHR, along with a third book (in 1980) of similar programs for the small computers then just appearing on the scene. That task completed, he left Iowa in 1980.

Classes on the use of computers have become perhaps the most obvious curriculum changes of the late 1900s. In 1968, two new classes (Analog and Digital Techniques for Data Reduction, and Numerical Calculations) taught students how to use computers to analyze data and to solve problems. While new classes of the 1970s addressed the environmental themes then attracting attention, in the 1980s, computers again gained prominence through additions like Computational Fluid Dynamics, Computational Hydraulics, and Environmental Systems Modeling.

In discussions of a new fluid-mechanics teaching laboratory, computers are being considered both as tools and as subject matter. In plans for this lab, computers programmed to simulate flows and processes will sit next to experimental models and traditional laboratory equipment. Students will be taught not only to use computerized equipment for data collection, monitoring experiments, and numerical modeling, but simultaneously to manipulate the computer models and the physical models in the "old-fashioned," hands-on manner.

As significant as these many impacts have been, they might be considered window dressing when compared to the influence computers have had on IIHR's research. Computers have revolutionized data collection, replacing the tedium of repeated on-site measurements with a flick of the switch. Long gone are the days when Joe Howe would trundle out to the weather station to manually record temperature, precipitation, wind, evaporation, and the appearance of the skies, teaching his son to drive along the way.

Data collection was the primary purpose of the IBM and remained a major task of the HP. By the mid-1990s, data collection had been relegated primarily to the relatively cheap, portable PCs that were abundantly scattered throughout IIHR's offices and annexes. For example, while data used to be fed from the basement towing tank to a centralized computer located four stories above the experiment, in 1997 a small PC perched on a carriage that glided up and down the towing tank above a ship model.

Computers are also routinely taken into the field to measure river profiles, and they might be attached to river-training structures to record details of their operation. The data-acquisition boards of "customized computers" allow large amounts of data to be fed automatically from the measuring instrument directly into the computer, where they are partially analyzed and stored. In addition, computers are increasingly interactive, meaning that they rapidly provide feedback to researchers and instruments, in some cases automatically adjusting the position of probes or changing data-collection parameters while an experiment is in progress. The development of "smart transducers" is taking this trend toward automation one step further. Microprocessors are being inserted directly into experimental measuring instruments, where they collect data, perform certain calculations, and feed data into the computers in semi-processed form.

While all of this ever-more-effortless data collection eases the lot of the researcher and increases the quantity of data collected, it can become a problem for the inexperienced. Somehow the act of taking a reading and writing it in a notebook challenges one to think about possible sampling errors. Mucking through the sediment in a flume or river develops a sense of the intricacies of natural systems, the eccentricities of models, and the difficulties of transposing these onto a sheet of paper. With the data-collection process becoming more and more automated and remote, students may never need to physically manipulate instruments or develop a sense of their short-comings or common weaknesses. If they first glimpse their data in partially processed form, neatly printed out in tables or graphs, they may never gain an intuitive, hands-on sense of the physics of fluid flow, and thus stand at risk of grossly misrepresenting the unpredictable complexities of nature. As Mark Wilson puts it, "Modern computers bring us closer than ever to the precise quantities we are measuring, but also can make us intellectually more distant."

Data processing is yet another story. The IBM could carry out very simple calculations such as averaging a set of numbers. But for the most part, data collected by the IBM were spilled onto heavy paper punch cards, boxes of which were lugged over to Weeg Computing Center to massage into useable form. In the 1980s, IIHR's acquisition of the HP allowed many more data to be processed on site, so that Weeg's large computers were tapped to a lesser degree. The arrival of UNIX workstations and ties to supercomputers have greatly magnified the ability to process data and perform numerical research directly from a researcher's office.

Thus, slide-rule calculations have given way to calculators, programmable calculators, and computers, the last two devices being increasingly pre-programmed to meet an individual's needs. By 1986, IIHR's Christmas letter was proclaiming that "the Institute's computer system is interconnected to practically everything in the lab, and clipboards and data pads are rapidly becoming obsolete." Each advance has allowed the researcher to address the same questions with greater depth and accuracy. Computers, in collaboration with steadily improving electronic measuring devices, have greatly speeded up and increased the accuracy of automated data collection, so that many more variables can be simultaneously recorded over a much larger area than before. Larger data sets can be more rapidly and easily reduced and analyzed, undergoing multiple iterations and being expressed in various graphical or tabular forms with the press of a key. These factors have increased the versatility and speed of research projects, and the number and sophistication of projects has shot up in response.

In addition to revolutionizing how research data are collected and processed, computers have shaped the very questions and

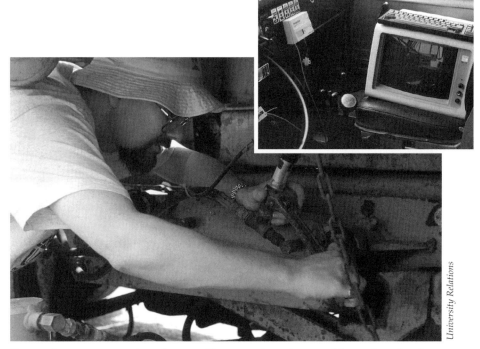

University Relations

Personal computers assigned to specific research projects and located at the site of the experiment are now scattered throughout IIHR's laboratories and field sites. Here, Wilfrid Nixon fits a snowplow blade with an electronic sensor that will feed data to a PC (inset), located inside the truck, that runs off the truck's battery. The sensors measure the force on the blade and its position, angle, and velocity, qualities important to Nixon's studies of ice control on highways.

types of research pursued, as well as the applicability of work done here to the world outside the university. In 1970, Jack Glover stated that "with the IBM 1801, we can conduct research that has been impractical, if not impossible, in the past." In the 1990s, this transformation is most evident among those who are performing numerical modeling in one form or another.

Numerical modeling refers to the use of computers to solve mathematical relations that govern the laws of fluid motion. These governing equations—for example, the Navier-Stokes equations and their idealized, frictionless forms (the Euler equations)—have been understood since the early nineteenth century. However, their numerical application requires a calculating ability far greater than the human brain can provide. Researchers have always attempted to apply whatever tools they had at hand to solve these nonlinear, partial-differential equations. But until the advent of digital computers, ever-more-powerful personal computers, and then workstations and supercomputers, the equations could only be applied to ideal fluids and solved in simplified form.

At IIHR, the initial forays into numerical modeling were made in the 1960s. However, not until the late 1980s were computers powerful enough to allow numerical three-dimensional modeling of complex flow processes of real fluids. This ability depended on two traits. First, computers became fast and powerful enough to handle tremendous volumes of data and thus provide the multiple iterations, in stepwise process, needed to solve the equations. Secondly, the graphics capabilities increased to the point that the solutions could be expressed visually, as a picture on the computer screen—a crucial feature, because the complexity and volume of calculations would not be understandable in tabular form. Color-coded depictions now allowed sufficient detail so that researchers could comprehend changes in flow, and moving visuals provided a picture of unstable flows changing over time. For the first time, the Navier-Stokes equations and their subsets, upon which any understanding of the physics of fluid flow depends, could now be solved in realistic situations.

This development has intensified the detail in which hydraulic processes can be examined and has allowed increasingly realistic graphical simulations of fluid motion that were previously unimaginable. In fact, numerical modeling may allow reexamination of questions with a depth that was previously unthinkable, and in doing so may provide new understandings of old problems. Numerical models, once perfected, promise to provide solutions to practical problems more cheaply and rapidly than laboratory models can provide. While Nagler watched with pride as fluids and particles flowed through his state-of-the-art transparent flumes, and Rouse proudly showed movies of these particles bouncing along the surface, researchers in the 1990s can in practice demand velocity and vector measurements of each individual particle in a sea or river of sediment. They can observe the flow of water around a ship's hull or trace the advance of a storm system via computer simulation.

How this type of knowledge will be applied remains open to the creativity of IIHR researchers. What is obvious is that numerical modeling (which includes computational hydraulics, computational fluid dynamics, and also the sophisticated climatological models of IIHR hydrometeorologists) is now being applied to nearly every field of endeavor at IIHR, from the tracing of water through a reservoir into and through a turbine and back into a river, to following the flow of air through a human larynx, to flows of water and air around a ship's hull and propeller. Numerical-modeling efforts provide a large percentage of research funding in the 1990s. They supplement continued studies based on physical modeling, forming a synergistic relationship between experimentation and computation, application and theory that generates a rich mixture of ideas and constitutes a major strength at IIHR. Thus, computer-based technology furnishes a new and powerful tool for understanding the complexities of fluid movement and, especially when enriched by traditional approaches to hydraulic research, it represents the frontier of IIHR's efforts.

15. River Engineering: Saving the Salmon

For thousands of years, they have followed the same pattern: emerging from their gravel-bedded nests, nibbling their way into fingerling status in their home waters, and then migrating down the Columbia River system to the ocean to mature while feeding in the high seas of the North Pacific. Pacific salmon are known for their complex life cycles, which depend on the integrity of cool, clear, freshwater streams in which to spawn and grow into smolts, as well as salty oceans that feed them into adulthood. After a few years at sea, the fish miraculously smell their way from distant oceans into rivers and up streams, back to the site of their birth, sometimes climbing over a mile in elevation and swimming over 900 miles upriver to reach their ancestral home. There, thick streams of fat-bellied females churn the water with their tails to form a gravelly nest or "redd" a foot deep, deposit their eggs, once again churn the water to fill the hole with gravel, and stand guard for a week or two before dying.

The masses of salmon that historically completed this life cycle each year have thrust this fish into symbolic status in the northwestern United States. Salmon are at the heart of Native American cultures, providing both sustenance and spiritual significance. For the white settlers who came later, salmon have become an economic mainstay, the focus of fishing and fish-watching, and an icon of the Northwest's wild character.

But while symbolic status remains, the salmon do not—at least not in the same numbers and strength as in past years. A hundred years ago, between 10 and 16 million salmon annually returned to the Columbia River. Major declines commenced early in the 1900s.

An extreme but not unique example is posed by the chinook salmon that return to the Columbia's tributary, the Snake River, to spawn in the spring and summer. This population numbered over 1.5 million in the 1800s. The number had dropped to 125,000 by the 1950s, to 59,000 by the 1960s, to below 10,000 in the 1980s, and to under 2,000 by 1994. This and other species of salmon have declined so precipitously that without intervention, they are expected to become extinct.

Multiple factors explain this tremendous decline: stream siltation, triggered by lumbering, mining, and grazing of adjacent hillsides, that smothers the eggs; warming of the water as shady streambanks are cleared; overharvest; predation by hatchery-bred fish; trampling by livestock; and other factors that have accompanied changes in land use.

Critical among these causes is the construction of dams and their consequent reservoirs for irrigation, hydroelectric power, and flood control. Rock Island Dam on the Columbia was completed in 1933; Bonneville in 1938; Grand Coulee in 1941; and between 1953 and 1975, 15 other major dams were built along the Columbia and Snake Rivers (with many more on their upper reaches) in an orgy meant to feed the craving for cheap electricity. By the end of the century, a mere 30% of the Columbia and Snake Rivers' water remained free-flowing in the 471-mile stretch winding its way eastward from the Pacific to the Idaho border. The remaining 70% slopped sluggishly against the dams and shores of reservoirs that made life hazardous for ocean-bound young salmon by, for example, doubling their migration time; subjecting them to increased predation and disease; and forcing them to pass through hydroelectric turbines or diversion structures. Returning adults also faced dam-related dangers.

Some of the earliest dams had blocked or flooded huge upstream salmon-production areas. Although concern about the salmon was expressed, lack of knowledge about the dams' effects and salmon ecology prevented the altering of structures to address declining fish numbers. As the dams proliferated, so did an understanding of the salmon and their needs. But not until the last few decades has knowledge progressed sufficiently to allow design of effective fish-passage systems.

One element of this concern was an attempt to assist the downstream-migrating juvenile fish past the many dams. Without any special structure, these smolts could pass the dams only by going through the turbines generating hydroelectric power, a passage whose turbulence and extreme pressure changes disoriented the smolts and exposed them to the turbine machinery and to predators downstream of the dam. The first diversions investigated were screens within the turbine intakes that guided the smolts into alternative channels through the dams. In 1982, IIHR received a contract to investigate these screens. This contract proved to be the start of major, continuing IIHR involvement in salmon-recovery studies, a focus that in the 1990s has netted the institute over a million contract dollars annually.

Similar screens had already been investigated for several years by the U.S. Army Corps of Engineers (COE), which had built the majority of Snake and Columbia River dams. The COE had developed a set of submersible, traveling screens that diverted the fish from the turbines into alternative channels, but the cost of maintaining and operating these screens was excessively high. Thus, IIHR's Jacob Odgaard, the project's principal investigator, started to develop a passive, non-moving screen system. The basic plan was straightforward: upon reaching the dam, the fish are guided instinctively by the water's flow down deep into the reservoir, toward the turbine inlet. There, a diversion screen intercepts their passage and sweeps the small fish into gate wells, where they are held in a "sanctuary box" until collected for downstream transport (or fed into a passageway that leads through the dam).

Although the system plan seemed self-evident, a number of variables necessitated that physical models of the setup be built and tested extensively at IIHR. The success of the models, built on a scale ranging from 1:4 to 1:100, depended on the accurate duplication of the river's flow conditions and fish behavior. Field studies of streambed topography, river velocity distribution, and the like allowed exact replication of flow conditions. However, the baby fish used in the studies (chosen for their small size) soon revealed a number of complexities. The velocity of water approaching the screen had to be precisely correct: if it was too fast, fish would be swept into the diversion screens, where they might be impinged or descaled and die. If the water was too slow, the fish would detect the approaching screen and swim below it into the turbine entrance.

Workers at the Priest Rapids Dam on the Columbia River prepare to install a screen designed to save the lives of young salmon by steering them away from turbine intakes. The screen's upward orientation will create a current that will sweep the small fish into alternative channels through the dam. An earlier version of this screen was tested through model studies at the institute (inset).

The water's velocity could be altered by adjusting the angle of the screen, by the size of its holes or bars, by their configuration, and even by whether they were left sharp or rounded off. Many tests of various screen sizes and configurations were made to determine the optimal combination of these features, the one that would capture the greatest number of fish with the least amount of danger to their fragile bodies. In some of the tests, water velocity around and through the screens was measured by lasers, which are able non-intrusively to make extremely precise measurements.

Additional tests were required to assure that the screens guided the currents (along with the baby fish) up toward the gatewells. The placement and length of the screens proved equally critical: too long and fish become impinged, too short and not enough fish are captured. And once the fish were swept up the channel toward the gatewells, the speed of the water had to decline gradually so that the fish were not stressed by turbulent currents within the gatewells.

The screens have been very successful in meeting their mission, and enough is now known about them that successful screen systems can be designed and commercially installed with relative ease. However, they remain tremendously expensive: a single screen costs about a million dollars, and one dam may require as many as 60 of them. IIHR has therefore been contracted to study other fish-diversion schemes, such as guiding fish into slots near the water surface above the turbine intakes, and then channeling them through a steel pipe in the dam to the downstream river reaches. These systems have the added advantage that the fish do not need to dive the 100 feet or so down (as they do when entering the turbine intakes), a rapid dive that submits their internal organs to the stresses of increased pressure.

Perhaps the simplest bypass would be allowing the smolts to pass over the dam's spillways. Such passage, although it costs the power companies something in lost water power, eliminates the need for diversion systems. However, it exposes the fish to an added danger: water supersaturated with nitrogen. Water foaming down a long concrete spillway plunges at the bottom into the river's depths, carrying along air bubbles that go into saturation as water depth and pressure increase. Fish following the bubbles down the spillway also plunge to the depths, where they absorb the dissolved gases. Gas pressure increases in their air bladders and goes from there into their blood, where the gases again form bubbles that can lead to death. IIHR has used its environmental flume, designed for modeling large, high-volume operations, to test various spillway structures that will prevent descending water (and fish) from plunging too deep into the river.

These two alternative diversion approaches have exemplified the need for precision models. The smolts by nature congregate near riverbanks, but these alternative passages require that smolts be guided instead toward the center of the dam's forebay, where the slots or spillways are located. Would a vee-shaped curtain device extending out into the forebay work in this situation? Because of the variables introduced both by fish behavior and by the nonuniformity of the forebay, various curtain models needed to be tested.

Measurements of water depth and speed were taken in the field, and then these were calibrated in 45-by-80-foot models, so precisely that small-scale adjustments were made by gluing individual rocks to the floor of the model. It is hoped that, with time, the results of these specific studies can be universalized to produce a general design for diversion curtains.

These studies have been both fascinating and challenging for several reasons. "They are truly interdisciplinary," states Odgaard, citing his team efforts with design consultants, power-plant clients, biologists, and others. "The research integrates the fields of fluid mechanics, environmental engineering, economics, politics, communications, and fish biology." A major consideration in the latter is what Odgaard calls fish psychology. "This presents a new mystery every day," he admits. "The fish drive the design."

The daily interaction with animals (rather than nonliving substances that behave repeatedly in a predictable manner) means that tests done one day may yield completely different results the next—and that researchers may be confronted with situations they could never have expected. "You think that you have the hydraulics all worked out, and build a model accordingly. Then you try it with fish, and everything needs to be changed because of variables you never could have anticipated," says Odgaard. Once, for example, the researchers built a model system with a square inlet pipe. But when fish were added, they did not pass through the pipe as expected. Instead, they clustered in the corners of the pipe, swimming upstream in the quieter water in a supposed attempt to control their passage rather than submitting to the faster currents of the inner pipe. The square pipe had to be replaced with a round one in which velocity was constant throughout. "This proved to be one small, unpredictable element in a model involving many such elements," admits Odgaard. "But it was a critically important element, one that had to work well for the total plan to operate. The whole project involved a composite of many such elements."

Researchers have been discovering how to get the young salmon over the dam. But what happens when they are released on the other side? There they face other dangers, for example predation by squawfish that lurk wherever smolts are released. The researchers knew that these predators could locate large concentrations of smolts by smell. Thus, squawfish congregated below the dam, where the turbines' churning outfall spewed out disoriented smolts along with the water. Obviously a better location for smolts passing through the diversion structures would need to be located.

To investigate this problem, IIHR researchers constructed two large models of the Columbia River below dams, complete with miniature diversion structures. Originally the smolts were to be discharged into a quiet pool about two miles downstream from the dam, but then researchers realized that squawfish found it easier to hold their own against the current in slower waters. Fearing that squawfish might congregate in the pool, IIHR researchers located a site 2,500 feet closer to the dam, where water ran rapidly enough to sweep the lurking squawfish away. This shorter outlet pipe saved at least a million dollars in pipeline-construction costs and increased the safety of passing fish, creating a winning solution for the company as well as the fish.

Other research focused on minimizing the depth of the plunging jet exiting the outfall pipe. Because squawfish lurked in deep waters, smolts would need to be discharged close to the river's surface, and with little disorientation so that they could best flee predators. Researchers started testing various outfall-pipe constructions. By discharging buoyant particles and dyes into model diversion pipes, the researchers could identify features such as the depth of the water's plunge and the spread of the outfall plume. They found that a simple pipe spat a forceful stream of water (and fish) downward to the river bottom—essentially feeding a concentrated flow of smolts directly into the mouths of squawfish. By testing numerous configurations for the mouth of the outfall pipe, they found that a spatulate-shaped pipe with an upward-sloping exit ramp would instead release a safely dispersed fan of fish gently onto the surface of the water, where predation was far less likely.

Through such studies, engineers are devising increasingly specific and effective methods for redressing the problems inherent in hydroelectric plants. Most of their knowledge has been gained through using physical models, which provide an accurate, economical, and relatively easy method of assessing complex river flows. The use of numerical CFD models may be even easier and far less expensive. However, the detail of CFD models historically has been inversely proportional to the size of the area being modeled: until the mid-1990s, only very small areas could be accurately modeled in three dimensions. In contrast, larger natural river reaches could be transcribed only as one-dimensional flows.

This is beginning to change. With the increasing capabilities of computers, the extreme complexities of simulating flows of larger river systems are being unraveled by institute researchers, who have created three-dimensional CFD models of specific river reaches near Columbia River dams. Two post-doctoral associates, Sanjiv Sinha and Ehab Meselhe, have succeeded in developing numerical models tracing water through the forebay of the Columbia's Wanapum Dam, into a turbine within that dam, and down the tailrace below the dam. "In the summer of 1996, for the first time, we took measurements on the Columbia to verify the numerical model of the Wanapum's forebay flow, and I was astounded at how closely the two correlated," claims Larry Weber, co–principal investigator on the salmon project. "Numerical models are advancing extremely rapidly. Our innovative work here is state-of-the-art for larger scale, three-dimensional CFD modeling more generally." Researchers hope to be able eventually to trace and manipulate the entire passage of water from its entrance into a reservoir through its discharge downstream, all on a computer screen. Numerical models have not been used much in making actual decisions regarding salmon management. However, once a model of a particular river reach is adequately calibrated and validated, it can be used for future testing of large-scale structural modifications to guide migrating salmon as well as for other purposes dependent on the river's flow. With the successful application of CFD techniques to the complex, irregular geometries found in natural rivers, along with their turbulent, unsteady, and stratified flows, hydraulicians will gain a sophisticated tool for tackling real-life hydraulic-engineering problems. Weber continues, "For years, our salmon studies have been shifting from qualitative to quantitative efforts, and the electric power utilities have been asking us for more and more detailed data. Gathering detailed data in the field, or even using a physical model in the laboratory, is just too expensive and time-consuming. But as the numerical models continue to improve, we will be able to use them to measure flow characteristics at countless points rapidly and economically, and make decisions accordingly."

A new 1:50-scale physical model of the forebay of Wanapum Dam was constructed during the summer and autumn of 1996. It

Today's advances in salmon studies rely on the interplay between field studies and research with physical and computational models. Studies of the Wanapum Dam (top) have necessitated the construction of one of the largest hydraulic models in existence, located at IIHR's Oakdale Annex #2 (center). This 1:50 scale model is being used to calibrate and validate IIHR's numerical models (bottom), which trace water flow into and through the dam. These constitute some of the very first three-dimensional numerical models of flow in natural, complex river sections.

occupies the newest IIHR laboratory, the spacious Oakdale Annex #2, and constitutes IIHR's biggest hydraulic model to date and one of the largest hydraulic models ever used for research, covering 8,000 square feet and holding 170,000 gallons of water. This model will be used to calibrate and verify the numerical models under development, and also to perform continued experimental work on this reach of the river and on fish-diversion structures that might be constructed there.

The numerical models still do not take into account the exigencies of fish behavior, which is even less predictable and more poorly understood than the meanders of fluids. But they potentially provide a powerful new tool for applying advanced fluid dynamics to fish safety and other real-life hydraulic and environmental flow problems. In 1997, fish passing over spillways or using bypass systems have survival rates of 97–98%, as compared with 85–90% survival of those passing by turbines. The structures designed at IIHR will guide more fish away from the turbines and thus will further improve salmon survival. Although this is but one factor in a broad attempt to revive the Columbia's salmon populations, improved bypass systems already have aided the cause by significantly improving downstream passage for salmon smolts.

Like so much research in process at IIHR, the salmon studies fit into a tradition of similar earlier efforts. In the 1970s, discoveries of fish impingement in the cooling water intake structures of the Quad Cities Nuclear Power Plant spurred the biological monitoring and remediation efforts associated with that project. Earlier still, in 1937, just before World War II, the Iowa Conservation Commission (now the Iowa Department of Natural Resources) engaged institute researchers in determining the most efficient system for upstream fish migration past Iowa's then-194 inland dams. The institute tested a number of fishways both through model studies and in the field—including tests of fishways for the Burlington Street dam on the Iowa River just outside the Hydraulics Laboratory. Investigations then (as now) included running live fish through scale models in the lab. The new fishway that was developed at the institute was hailed by the press as highly successful because "fish emphatically do not like to bump their heads," a problem eliminated by the new fishway. A model of the new fishway was even taken to the Iowa State Fair as one of the university's success stories for 1941.

IIHR's fish studies began shortly before World War II, when researchers developed a new fishway that allowed fish to negotiate Iowa's small dams. The new fishway, tested with live fish, provided a slower, smoother passage than traditional fishways.

Several years before this, in 1932, fish-passage systems had been on the minds of IIHR researchers. On July 11, Floyd Nagler wrote a letter to Jacob Crane Jr., the planning consultant who was then co-authoring *The Iowa Twenty-Five Year Conservation Plan* (a landmark document that reshaped the state government's management of Iowa's natural resources). Nagler wanted to protest the "utility of fish-ways on Iowa dams." His arguments against them were multiple: they leaked pond water, they were annoying to mill operators, fishing would improve without the passageways, and they were (in his estimation) not needed by the fish. However, he closed the letter by admitting that he knew little of fish life and was merely raising questions about the advisability of fishways. In conclusion, he conceded, "If the Department thinks that real benefit to fish life would be secured by allowing the free passage of fish over the dam, I am convinced that considerable thought should be given to the design of a type of fish-way that will actually accomplish this purpose." We can only wonder what Nagler might have thought had he realized that in writing this, he was scripting a charge that would be taken up a half-century later by researchers in his own laboratory. And that these researchers, receiving millions of dollars of research funding, would struggle with the challenges presented by gargantuan dams that made Iowa dams look like toothpicks and threatened the very existence of one of the nation's significant fish groups.

16. Hydraulic Structures: Dropshafts beneath Our Cities

Vast networks of city sewers form a dank and dark underworld inhabited by imaginary Ninja turtles along with scattered raccoons, Norway rats, profuse microorganisms, and other real-life residents. Webs of coalescing conduits carry away the sewer waters that have washed our bodies, clothes, dishes, and floors, and serviced our industries. In most older cities, these waters mix with the storm runoff that has flowed across our streets and parking lots. The mesh of pipes located five to ten feet underground captures and channels the effluents that we wish neither to see nor to think about, pumping them onward out of sight, out of mind, and out of our lives. All runs downhill in a watery matrix into ever-larger pipes that eventually lead to water-treatment plants. Cleansed, the water is then released back into the natural world.

These are the storm-sewer systems that make life in the modern city possible. They function well to rid us of our liquid wastes—the grime from sinks and showers, dissolved food from dishwashers, unwanted goldfish and everything else flushed down toilets, industrial effluents, and the rainfall that washes our streets and carries away leaked oil, spilled gasoline, and the grit that veils the surface of the modern city. Simple, gravity-fed devices, these systems allow human life to proliferate in complex, densely packed living quarters, without concern for what will happen to the quantities of street-dirtied precipitation or waterborne wastes produced in the city.

That is, the storm-sewer systems did so until a few decades ago, when they faltered under the stress of growth pangs. The largest cities were expanding beyond the wildest dreams of their founding fathers. Having nowhere else to go, people crowded into smaller and ever denser areas by going up into high-rise office and apartment buildings. With more people on less land, the storm sewers were being asked to handle ever-larger quantities of sewer wastewater. At the same time, rainwater runoff was becoming more problematic. With even the smallest city plots being paved and put to use, precipitation was unable to percolate into the soil and instead was increasingly forced into the streets and storm-sewer systems. Thus, the pipelines were taxed with ever-increasing surges of both storm and sewer water. Sometimes, when rains were especially heavy and water treatment plants became overtaxed, a torrent of wastewater was released into rivers or lakes untreated. With rising environmental sensitivities and the passage of the Clean Water Act in 1972, such periodic flushing of raw wastewater was no longer acceptable.

Clearly something had to be done. Water exceeding the capacity of the storm sewers would have to be held someplace until the treatment plants could handle it, but where? With the ground surface of large cities already fully occupied and property values skyrocketing, the only place to go was underground. When Milwaukee started to plan an underground storage system in the early 1980s, the city came to IIHR to request help with avoiding the mistakes that its neighbor to the south, Chicago, had made ten years earlier. Chicago had been one of the first cities in this country to plan a massive underground tunnel and reservoir system to store excess storm-sewer water temporarily. The system was named TARP, the Tunnel and Reservoir Plan. The idea was to run a few gigantic conduits, up to 36 feet wide, deep under the city. These would connect five underground football field–sized reservoirs big enough to handle the flows associated with even the largest imaginable thunderstorms. The wastewater in the system would be pumped slowly into treatment plants as space became available. Because the cost of constructing and laying massive preformed conduits would be unimaginable, the lines would simply be bored into the bedrock beneath the city. The interface between the near-surface storm-sewer system and the deep storage conduits would be a series of vertical pipes called dropshafts.

Chicago's engineers had quickly realized that a closed system of this size, carrying fast-moving rivers of waste deep underground, can pose many problems. The water falling vertically down a long dropshaft is noisy, and it is forceful enough to structurally damage the tunnel bottom. Foul-smelling air entrapped in the conduits could "burp" back up a dropshaft into city streets and propel the dropshaft cover skyward. Surge waves could send liquid spouting into the streets. Changing pressures inside the underground tunnels could turn water into vapor one minute, and a moment later collapse the vapor bubbles with a force strong enough to pit the conduit walls. When Chicago's planned system was tested with small-scale models, another problem revealed itself. The air entrained in the water plunging down the dropshafts not only took up too much space, it caused major hydraulic instabilities that resulted in massive vibrations that some feared would structurally damage the dropshafts.

In an attempt to alleviate these problems, the design of Chicago's dropshafts became increasingly complicated, intricate, and expensive. A plunge pool was added at the base of the dropshaft to neutralize the force of falling water. The roof of the conduit next to the base of the dropshaft was slanted upward to create a "deaeration chamber," where any remaining entrained air would be collected and removed. A perforated partition was inserted to divide the dropshaft in half, creating one chamber for the water to go down and a second parallel chamber for air to escape upward. The complexity and cost of TARP rose higher and higher until the multi-billion-dollar system became a political and fiscal fiasco.

Milwaukee realized that it needed a similar gargantuan underground conduit system, with dropshafts 4 to 13 feet wide plunging as much as 300 feet downward to the storage conduits, but it did not need Chicago's headaches. The city engineers came to IIHR to commission a different set of engineering designs. "We knew nothing about dropshafts at the time," confessed Subhash Jain, the professor who became the institute's dropshaft expert. "But it was a challenge, and we thought that we could handle it. So we assured the Milwaukee engineers that we could improve the Chicago system, and we got to work."

The main goal was to construct a smoothly operating structure as cheaply and easily as possible. That meant a straight cylindrical dropshaft of uniform diameter, attached in a simple right-angled juncture to the deep uniform horizontal conduit. Each line could be bored by tunneling with a single round cutting head. How would the dropshaft operate? IIHR researchers knew that a vortex flow would avoid many of the problems experienced in Chicago. Water swirling in a vortex around the walls of the dropshaft would leave a central column of air in the dropshaft. The rapidly moving water would entrain less air than water plunging straight down, and the central air column would vent some of the trapped air back up to the atmosphere. A deaeration chamber at the bottom would remove the rest. In addition, the vortexing water would dissipate much of its energy through friction with the dropshaft walls. Thus, structural damage at the shaft bottom, and the need for additional constructs such as plunge pools, would be avoided.

But how could the water be set to swirling? Complicated inlets built in a spiraling or scrolling configuration had been developed elsewhere—the "orange peel," for example. But these complex inlets occupied too much space and were expensive and difficult to construct. IIHR developed the tangential inlet for Milwaukee, a simple descending ramp that shot the water into the dropshaft at an angle and thus sent it twisting rapidly around the dropshaft walls. This tangential inlet, which took far less space and was cheaper and easier to construct than other vortex-flow inlets, was a major breakthrough.

The inquiries and contracts of other cities pushed IIHR to further its dropshaft research. Phoenix, Arizona, decided that its needs could best be met by laying smaller-sized storage conduits in trenches cut into the ground surface (rather than boring the conduits into the ground), but the space requirements and associated costs of deaeration chambers would be prohibitively high. The city asked IIHR to develop a dropshaft that completely eliminated the need for these chambers. The "helicoidal ramp" was the solution—the walls of the dropshaft would have a set of internal twirling vanes added, that would keep the water pivoting rapidly from the top to the very bottom of the shaft so that air would not become entrained and thus would not need to be removed. The tangential inlet was retained.

Once the helicoidal ramp concept was developed, further research revealed that the vanes need not extend the entire length of the dropshaft. A "truncated helicoidal ramp" with vanes only near the bottom would be sufficient. Dropshaft construction is expensive; any elimination of detail was a money saver. The world's largest city, Tokyo, approached IIHR with a request to adapt this truncated ramp

IIHR researchers demonstrate a model dropshaft system to visiting Milwaukee engineers. The water plunges down the larger vertical shaft in the forefront into the horizontal deaeration chamber. Air mixed with the water is vented back up to the atmosphere through the smaller vertical pipe, while the water continues to flow forward through the underground storage conduits. Note that all components of the dropshaft system remain round and straight to minimize underground tunneling costs.

system for its needs. In addition, Tokyo wanted a replacement for the tangential inlet into the dropshaft. In Tokyo, where the land's surface is so valuable that even large rivers are shunted underground through tunnels, there was no room for the above-ground construction operations that such an inlet required. IIHR came up with yet another dropshaft structure: a helicoidal ramp that was truncated but incorporated vanes to swirl the water at the very top and very bottom of the dropshaft. The water would enter the dropshaft through a simple radially connected round pipe, which could be constructed through tunneling. Water would be set swirling around the outside of the dropshaft by the uppermost vanes. Then the water would enter a central, straight, vaneless section of pipe, in which

IIHR researcher Subhash Jain (on ladder), here assisted by M. P. Cherian, developed a system for feeding water into a dropshaft through a tangential inlet, which sent the water swirling around the outside of the dropshaft and thus created a more stable downward flow.

momentum would keep it rotating but at a slower and slower rate. Vanes near the very bottom of the dropshaft would speed up the swirls once so that entrained air would be released into the open core of the dropshaft, and no deaeration chambers would be required.

Today, several of our largest cities have installed mammoth underground water-storage systems using IIHR's designs for their structures. When special needs arise, IIHR is still contacted for modifications of the basic plans.

Dropshaft studies, like much current research, has roots that reach deep into IIHR's history. One of the institute's first large experimental projects was a series of tests on the flow of water through culverts. In the 1920s, over 3,000 tests were made on various culverts by Nagler along with his cohorts Yarnell and Woodward. The results were described in the institute's first bulletin. Starting in the 1930s and continuing for decades, extensive studies were conducted using pipe systems that were torn down and refigured for new efforts. The hydraulic jump, the mixing of air and water, characteristics of flow at sudden and gradual expansions in a pipeline, and laminar and turbulent flow were all studied in pipe loops and conduits, along with several other topics: sediment transport in pipes, converging and diverging flows, motion around bends, pressure and velocity at outlets. These provided a rich mixture of topics for graduate theses, as well as developing a tradition and basis for the dropshaft studies.

Dropshafts fit into IIHR's larger research picture in additional ways. They are but one of the many hydraulic structures shaping the flow of water that have been investigated here, others being locks, dams, spillways, sewer systems, salmon-passing and other river-training structures, pump intake structures, and cooling towers. Research on dropshafts has become one of the legacies of the Kennedy years, with their expansion into applied areas of research and their emphasis on environmental projects. Dropshaft projects have also been a single facet of IIHR's urban hydraulic research, which focuses on the safe and efficient provision, transport, and processing of the fluid most crucial to human life—water—in the densely populated cities that are now inhabited by the majority of Earth's people.

17. Cold Regions Engineering: Arctic, River, and Road Ice

Nome is the principal settlement of the western Alaskan coast, despite its population of only 3,500 people. One might wonder why anyone would choose to live there, where the winter sun fails to rise above the horizon for nearly two months of the year and the frost-free summer lasts only a few weeks longer. Nome's temperatures drop below freezing for two out of every three days of the year. The harbor freezes for seven months of the year, leaving the town icebound. Winds of winter storms sweep in from the Bering Sea, blowing across Norton Sound at 70 miles per hour, pummeling the village and shepherding churning masses of sea ice a yard or more thick onto the shore.

In the mid-1970s, Nome proposed constructing a major port facility in this harsh environment. The nation's demand for oil provided the impetus. Alaska's "black gold" promised unlimited wealth. By 1977, massive engineering activities were helping to convey crude oil from the nation's largest oil field at Alaska's North Slope through a newly constructed 800-mile-long pipeline crossing the vast northern wilderness. Logistical support was needed for North Slope drilling activities. Nome seemed to be a likely staging point for ships carrying drilling equipment and supplies, and so the city decided to construct a port that could handle large transport ships. The port also could service Nome, which until then had received its supplies from barges that shuttled between Nome's small, shallow harbor and supply ships moored several thousand feet offshore.

Plans called for a causeway that would stick two-thirds of a mile straight into the sea, leading to an offshore terminal and docks planted in water deep enough for seafaring ships. With this facility, goods could be unloaded from ships directly onto the wharves. With time, two additional piers could be added for servicing rigs and ships involved in oil and gas exploration.

Although it appeared that the causeway, armored by 20-ton boulders to protect it from the sea's waves and ice floes, would provide a safe haven for ships, questions remained regarding the causeway's ability to withstand winter's wind-driven ice packs. Would Norton Bay's packs mount the causeway? If so, what design features could minimize their damage? And how could the floes of pack ice shoving through the ships' approach lanes be managed?

Because of their expertise in ice engineering and modeling, IIHR's John Kennedy and Robert Ettema were approached for answers to these questions. In addition to visiting Nome to inspect the proposed construction site and assess the characteristics of Norton Sound ice, they studied the interactions of ice floes and the causeway with hydraulic models. The studies yielded useful insights. The first wave of ice assaulting the causeway left a protective blanket of frozen rubble that freed the causeway from future ice damage. Provided that the causeway and road shoulders near the bridge were sufficiently substantial, they would be able to withstand the assaults of winter. However, the ice could override the causeway. While some ice might continue over the causeway and reenter the sea on the other side, some might remain to block access to the terminal ports. The overriding of ice could be inhibited by placing massive tubular ice-breaking frames in the water, on the windward side of the causeway. These frames would render the port useful year-round if icebreakers were employed to clear the port area and if heavy machinery regularly removed the ice from the causeway and docks.

Ironically, shortly after the study was completed, the price of crude oil plummeted, and much northern oil exploration was discontinued. Nome's plans for expansion shrank. A simpler form of the proposed pier has been constructed, a robust causeway extending into deep water that is kept open only during the five ice-free months.

IIHR's first ice-research projects had been undertaken nearly 15 years before the Nome project, when ice engineering had been a

(Top) A model of Nome's proposed harbor in miniature was constructed in Iowa City, a thousand miles from any ocean, complete with a roadway, flowing ice, and model cars. This model revealed that wind-driven ice floes could mount the proposed causeway and block it during the winter.

(Bottom) Photos taken on a jetty at the entrance to Nome's harbor show the ice piling up in a manner very similar to that demonstrated by IIHR's model studies. The jetty is a small version of the causeway subsequently built at Nome.

brand new and unexplored research field. In the intervening time period, much of IIHR's effort had been devoted to understanding the complex properties of ice as a material (for example, its formation and strength under varying temperatures and pressures), as well as deciphering its destructive characteristics and its effects on river flow. Major research topics had included the formation and behavior of frazil ice, fine ice crystals that form in supercooled water. Frazil adheres to and clogs inlet pipes, locks and dams, and other hydraulic structures and also forms problematic ice floes and hanging dams. Other research topics included ice forces on structures; mechanical properties of ice, such as its bending and crushing strength; dynamics of ship hulls moving through ice; and flow in ice-covered streams, including topics as esoteric as ripples on the underside of ice sheets. The institute's studies had included a mixture of the theoretical and applied. Ice, a complex substance about which little was known, needed to be better understood before its damage could be controlled.

Cold-regions research had been stimulated by the push to develop the earth's northern regions and especially to mine their resources. In addition to the Nome study, IIHR had investigated the dynamics of floating ice sheets and the effects of ice collisions on offshore oil platforms. After the mid-1980s drop in oil prices had collapsed oil companies' interest in funding ice research, ice problems elsewhere continued to bring contract funding to IIHR. Some projects concerned attempts to free large rivers—the St. Lawrence Seaway, Upper Mississippi, Ohio, and Illinois—from ice-induced shipping disruptions, damage to locks and dams, and flooding. Keeping these channels open to year-round navigation is a matter of tremendous economic importance. One IIHR study examined the possibility of using thermal effluents for suppressing ice formation; another, the effects of frequent passage of boats on the refreezing of a river channel; a third, using unconventional energy sources to de-ice locks and dams.

Ice jams on rivers have been a concern in the Midwest and a consistent research emphasis at IIHR for the last quarter century. Ice forming in the fall or breaking up in the spring usually flows down rivers unimpeded. But should these ice floes meet an obstruction, they pile up to block both water and ice. Such ice jams form rapidly and without forewarning. Sometimes they cause devastating floods. Later, when a jam fails, the fast-moving ice may severely damage structures along river channels and on their floodplains. Over the years, attempts had been made to control ice jams by blasting them with dynamite, a practice that has succeeded infrequently. IIHR's formulations and computer models have been successful in developing a fundamental understanding of ice jams and thus helpful in predicting their formation and effect on water levels. Research that began in 1995, on ice jams at the confluence of the Mississippi and Missouri Rivers, seeks to decrease the incidence of these navigation-blocking jams through altering channel morphology and the rivers' flow distribution.

Many of IIHR's ice studies in the 1990s focus on winter travel hazards, a major interest of researcher Wilfrid Nixon. Ice on roads is controlled primarily with salt and sand, both of which can cause environmental problems. Salt damages vehicles, the road surface, and the plants and animals that are exposed later to salt-contaminated water; sand can degrade air quality. The Iowa Department of Transportation (IDOT) has been funding a series of research projects aimed at increasing the efficiency of trucks scraping ice from the roads. Nixon uses a precisely calibrated high-speed ice scraper to test the effects of various cutting edges moving across ice surfaces. Particularly effective blade designs are tested on the road, by research trucks fitted with underbody or front-mounted plows and with pressure-monitoring and data-logging devices (see figure on page 213). The performance of blades passing these tests is then tested in the field on IDOT trucks. Nixon has concluded that a sharp blade with a single point of road contact is far more effective in clearing ice than is a heftier, broad-surfaced blade. In related research, Nixon has been investigating the strength and deformation properties of ice, especially as they relate to the mathematically complex characteristics of ice adhesion to a variety of surfaces and the formation of cracks in ice.

Nixon's research is carried out in part in IIHR's extensive ice facilities, the five refrigerated chambers located in the East Annex. In two of these, specific types of ice are grown under closely controlled temperatures and conditions. Very cold temperatures, down to -45°C, are obtained in a third room where fundamental research on ice as a material is performed. Here, strength and deformation properties of ice are tested under a variety of load conditions.

The fourth cold room contains equipment for preparation of ice samples and Nixon's ice scraper. The final chamber contains a refrigerated towing tank. Here a 21-meter-long towing tank can be kept at temperatures down to -20°C. Ice-vessel and ice-structure interactions can be tested by using the tank's carriage to tow models through the ice-rimmed water. Patterns of ice fracture can be observed from all angles: in this tank, videos can be made by an underwater camera, a feature that allows ice formation and break-up to be observed simultaneously from above and below.

IIHR's ice laboratory also includes an icing wind tunnel, used for studying the deposition of ice on cables and larger structures. This is of great importance, for example, in the transmission of electricity; the weight of ice deposited on electrical transmission lines and towers can snap and topple these structures. These laboratory instruments and facilities allow IIHR to claim status as one of the few comprehensively equipped and staffed university-based ice laboratories in the world and as the training site of many of this country's ice engineers.

18. Hydrometeorology: Storms over the Midwest

Facing north on a blustery Iowa fall day, one can almost smell the winter blowing from the Pacific Northwest across the plains. Iowa's flatlands offer little protection from the promise of the stinging bite of wind-driven crystals. Soon snowstorms will rage. Sleet will blind cattle, and school buses will skate over ice-coated county roads that send cars careening into drift-hidden fences. Pioneer stories tell of Midwestern farmers getting lost on the way to the barn, their bodies being found after the thaw.

In the summer, warm masses pregnant with moisture from the Gulf of Mexico billow up from the south. Thunderstorms attack with the ferocity of a tiger. Forks of lightning blaze through black skies 40 miles to the west. Hailstones damage a year's worth of corn, and a girl fearing a tornado huddles with her cat in the corner of a basement. The Mississippi River, oblivious to the towns lined up in rank on either bank, surges above its levees once again.

It has been said that Iowans obsess on the weather. If true, perhaps it's for good reason, here where weather remains our strongest tie to the prairie wilderness that once was. The skies remain open from horizon to horizon, undaunted by mountains or gorges. Winds bringing in air masses laden with moisture can roam freely in Iowa. We examine these events as elements of chance, with limited predictability and unlimited power to create or destroy our structured lives. The news continually forecasts what weather to expect, but we know that what comes will come, whether predicted or not. No one tells the weather how to behave in Iowa.

In a secluded lab at IIHR, Witold Krajewski checks with a graduate student who has just pulled up a computer image, generated through radar data, of a storm system over northern Australia. Krajewski is a hydrometeorologist, someone who works with the science of moisture moving through the skies. He and the student play with the colors, deepening the reds to reflect the highest intensity of what appears to be a giant storm system dumping inches of rain during the monsoon season. The two debate details of a complex equation designed to help convert the abstract image, the reflection of energy off rain or snow particles floating through the air, into accurate estimates of the quantity of precipitation falling per unit of time. Satisfied that their algorithm will yield a rainfall measurement consistent with a rain gage, they close the file. Before leaving for the evening, Krajewski and his research colleague Anton Kruger make one last visit to the rooftop of a nearby building to check their weather instruments. It has been a stormy spring here in Iowa. The disdrometer constructed to measure the size and shape of raindrops has just been installed and seems to be battened down securely. It is now ready to feed data into the computer system downstairs. When they leave, the skies look clear, although a low line of clouds is building in the west. That night a storm dumps six inches of rain, sending creeks surging out of their beds to flood fields for miles around.

"It's sometimes hard to explain the connection between our work and the weather events that are so important," explains Krajewski the next day. "Our efforts seem too obscure and academic to many. Yet the connection is strong. The TV weatherman looks at a radar image of storms moving across the plains and speculates that there will be rain the next afternoon. Farmers, vacationers, managers of Iowa's reservoirs, they all shape their lives around those very rough predictions, which are totally nonquantitative. Here at IIHR, we are trying to use the same types of remote sensing radar devices to develop a model that will accurately predict not just the occurrence of rainfall, but also the exact quantity of precipitation that will drop in a specific location. The implications for forecasting extreme events and water-resources management are astounding."

Two of Krajewski's research projects help explain his efforts. The first, funded by the U.S. National Weather Service, attempts to help that branch of the government to link its new radars to complex forecasting programs. The Weather Service in the 1990s has replaced its entire radar system with WSR-88D instruments that are much

Photographs taken in 1918 (top) and 1993 (bottom) at the site of the Hydraulics Laboratory show many changes in the university's Power Plant (directly across the river) and downtown Iowa City. However, there is very little difference in the floodwaters of the Iowa River, whose flow has obliterated the Burlington Street dam (see page 22 for a view of this dam during normal flow). IIHR hydrometeorologists are working toward minimizing the damage caused by such extreme events by developing accurate predictive precipitation models.

more precise, and operate on a much smaller grid, than the radar system previously used. The radar yields an image of precipitation and storm systems. The radar image is constructed with the reflected energy of microwaves being bounced off the rain droplets: many or large droplets and the reflection is great, small or few droplets and the reflection declines. Radar images, when coupled with modern computers, have the hypothetical capability of yielding rain measurements. That is, the qualitative images could be transformed into "real-time" quantitative information accurate throughout an entire region. Weather forecasts then could integrate these remotely collected radar data with those from on-site rain gages, which tell the Weather Service about precipitation only after the fact and at a single point. However, the multi-stepped mathematical conversion is very complex, and inaccuracies are introduced at every step of the process. Discovering the source of introduced errors and attempting to limit the inherent uncertainty is at the crux of Krajewski's work, as is working with the concept of the limits of weather predictability.

In the mid-1990s, an hour's worth of radar images could be converted into precipitation measurements that could be verified by rain gages. Krajewski's immediate research goal is to increase the conversion accuracy until a single radar image, collected instantaneously, can be immediately translated into a precipitation estimate. His ultimate goal is to apply that capability to the future: to develop the ability of using the radar images to predict future precipitation events and quantities. Improving the accuracy and timeliness of such instantaneous assessments of rainfall over large areas would be extremely helpful in issuing warnings about extreme weather, such as stream flows and floods. The accuracy of weather forecasts, stream flows, and other weather-related environmental responses could be improved, as could the management of water resources, for example the release of water from a reservoir or a water-treatment plant. These data would also allow the development of a new generation of hydrologic and climate models, including models of global air circulation and global climate change.

To accomplish his ends in this project, Krajewski uses a sophisticated computer laboratory and a weather station on the roof of the East Annex building, both of which are kept in working order by Kruger. In addition to the normal weather instruments that measure

Anton Kruger poses in front of IIHR's optical disdrometer, a sophisticated instrument that measures the distribution of raindrop sizes. Such measurements are crucial to understanding the accuracy of radar-generated rainfall estimates, since the scattering of radar impulses is affected by raindrop size and shape.

humidity, wind speed and direction, temperature, and rainfall with both tipping bucket and optical devices, this station includes two special tools: a small radar dish that points straight up to display the vertical structure of rainfall and a newly installed optical disdrometer (see figure, page 245). The dish records reflected radar pulses along 10- to 15-meter vertical intervals. This high spatial resolution is matched by a high temporal resolution—the radar (and other weather instruments) record data every two seconds, compared to the Weather Service's typical five-minute interval. Additional instruments are being added as allowed by new grants and are being implemented at the Iowa City Airport and on the Mobile Rainfall Observatory, an instrument-fitted trailer that can be moved to record weather variables in different locations. Together, the complex of unusually sensitive instruments yields a very detailed pictorial representation of storm systems. Few other universities have an equally sensitive collection of rainfall-measurement research instruments.

The second project, funded by NASA and named the Tropical Rainfall Measurement Mission or TRMM, is a multinational initiative examining the climatic belt surrounding the earth's equator, a region whose climate is poorly understood. The ultimate goal is to be able to assess rainfall there remotely, through satellite-borne radar. TRMM, which focuses on tropical areas where weather stations are relatively few and far between, thus suggests an alternative to the ground-based radar systems used in temperate climates where weather stations are more abundant. IIHR is helping to evaluate the algorithms that will translate the satellite radar data into actual precipitation estimates and will test the validity of that translation by relating these data to ground precipitation measurements. Once these algorithms are developed, the long-term study will test them by applying them to radar data on rainfall from a low-flying satellite, launched jointly by the United States and Japan, that will circle the equator. The satellite data will be validated by ground-based radar data, which in turn will be validated by ground-based weather stations and their rain gages.

Knowledge about the tropics is especially important if we are to trace the climatic changes that the tropics may experience as their human populations continue to swell and their rainforests decline. Rainforests shape tropical climates through returning to the atmosphere (through transpiration and evaporation) a significant amount of the precipitation falling on them; without the forests, the climate of the region could become significantly drier. Remote satellite radar data are likely to remain an important precipitation information source in the tropics, where there are few rain gages and relatively small land surfaces.

But these studies are of greater than regional importance. For one thing, most of the earth's rain falls on the tropics. The tropics are regions of intensive atmospheric energy exchanges that shape air circulation patterns around the globe and strongly influence the climate in temperate regions around the world. For another, rainfall data are a way of measuring an entity called the latent heat budget. The sun's energy is used in part to convert water on the earth's surface into clouds that are blown to distant locations. Eventually the moisture condenses—releasing its translocated energy in the form of heat—and falls as rain. Latent heat is a crucial factor in global circulation, simulated in climate-change models that are extremely important in deciphering the long-term effects of our society's atmospheric wastes. Because latent heat is extremely difficult to measure directly, researchers derive their figures indirectly from rainfall measurements. "That's the amazing thing about this research," states Krajewski. "Through a complex series of extrapolations, we are studying the size and shape of raindrops here in Iowa so that we can predict the climate of the entire planet for the next 50 years, and develop models that will work with concepts of global climate change."

Krajewski fits into a tradition of hydrological research that has characterized the Hydraulics Laboratory since its inception. One of Floyd Nagler's first projects, commenced in 1922, had been the establishment of monitoring stations for rainfall and runoff throughout the Ralston Creek drainage just east of Iowa City. The Rapid Creek watershed was added to the project in 1938. This monitoring project resulted in a data set heralded as the longest highly detailed record of a small U.S. watershed in existence. The Ralston Creek surveys were continued until 1988, becoming IIHR's longest-lasting continuous effort. Their data have been used by numerous students and senior researchers.

Nagler also wrote numerous reports describing Iowa's streams and rivers. His interest in hydrology—the flow of water over and through the ground surface—was reflected by the stream of hydrologists who followed, such as Joe Howe (with a record 42-year tenure at IIHR that began in 1929) and Tom Croley in the 1970s, followed by Peter Kitanidis in the early 1980s, and in 1997 represented by the groundwater hydrologist You-Kuan Zhang, among others.

Joe Howe, in addition to supervising the monitoring projects on Ralston and Rapid Creeks, directed the university's voluntary collection of traditional Iowa City weather data for the U.S. Weather Bureau. He started the practice in 1937, when the Iowa City station was heralded as one of the best in the state because it had an evaporation pan (in addition to the usual thermometers, rain gage, and anemometer). Howe used the daily drives to the equipment site near the stadium to teach his son how to drive and continued these data-collection efforts until he was 77.

IIHR's more recent emphasis on the atmospheric flow of moisture, hydrometeorology, commenced during the years that Konstantine Georgakakos was at the institute (1985–94). His studies incorporated new weather instrumentation and computer modeling techniques into flood and flash-flood forecasting. For a short period in the early 1990s, when Professors Ignacio Rodriguez-Iturbe and Rafael Bras were present, IIHR entertained what was probably the strongest hydrometeorology research group in the world.

The diversity of hydrological studies that has characterized IIHR's activities continues to be embodied by IIHR researchers. Allen Bradley brings a variety of research perspectives to his interest in the hydrometeorology of extreme events (floods and droughts). Bradley's research has run the gamut, from characterization of past drought frequency by using tree-ring data, to analysis of the Midwestern floods of 1993, to studying trends in extreme rainfall events of the twentieth century for signs of more massive regional and global climate change. He believes that the small-scale analyses possible with the NEXRAD system will have multiple operational and planning uses—such as predictions of stream flows throughout a watershed, and provision of detailed rainfall and soil-moisture maps to farmers during the growing season.

Another IIHR faculty member, Frank Weirich, as a geomorphologist mixes questions about the effects of running water into integrated watershed studies. Investigating water as an agent of landscape change, he traces precipitation from the moment it hits

IIHR's hydrological research dates back to early in the century, claiming a longevity that characterizes many subjects investigated here. Joe Howe (crouching, in suit) took charge of the institute's weather-data collection for over 40 years.

the ground and picks up sediment particles until the water enters a lake or reservoir and the sediment comes to rest. Weirich is especially involved in the roles of fire and flood as agents initiating change in landforms. Like Krajewski, he combines field studies with computer applications, in this case using GIS computer systems for full-scale landscape modeling.

Other additions to the permanent staff include Anton Kruger, who became an IIHR research scientist in 1996, and Bill Eichinger, who joined the faculty in 1997. Kruger has been in charge of IIHR's ultra-high-resolution surface meteorological station for several years and has also developed a variety of computer software programs for the hydrometeorology group. Notable efforts along these lines include development of a visualization software package called VRAD that the U.S. National Weather Service is considering for use with their NEXRAD system, and the organization of very large radar databases that require on-line storage of 100 Gbytes or more. This organization allows researchers to compile years' worth of radar data so that they can be visualized and analyzed in a meaningful manner. Kruger's work is crucial to IIHR's research on rainfall estimation from weather radars.

Eichinger, an experimentalist, studies atmospheric transport and surface-atmosphere processes through precise measurements of diverse atmospheric variables over fairly large areas (5 to 20 km^2). His measurements are made with laser-based instruments, tuned to carefully selected wavelengths, which send out a pulse of energy that reflects off the evaporating moisture, rainfall, or any other atmospheric constituent Eichinger may be measuring, from particulate air pollutants to particular types of molecules. The pulse's transit time relates to distance, which allows him to make two- or three-dimensional maps of various atmospheric constituents. The studies of pollutant particles expand Eichinger's work beyond weather investigations into research on environmental contaminants such as photochemical smog. He predicts that his lidar system, which is being applied even to measuring the wind, will provide the fastest and best-resolved wind sounder in existence. The expertise represented by Eichinger and other additions to the hydrometeorology group assures that IIHR's hydrological capabilities and interests will continue to expand to include an increasingly wide range of applications.

19. Ship Hydrodynamics: A Summary

Boats and ships have shaped human history for thousands of years. They have carried explorers to new continents, built new economies by transporting raw products and manufactured goods, borne warriors and weapons of defense as well as attack, opened the gates to mining the sea's fisheries, carried immigrants to new homelands, and engendered numerous forms of pleasure.

Until recently, either wind or human sweat propelled ships through the water. That changed significantly in the nineteenth century with the perfection of the propeller. Propellers, powered first by steam and later by internal-combustion engines or, still later, by nuclear reactors, pushed bigger ships forward at faster and steadier speeds and less cost than the wind drove the clippers of the day. Growth in the size of ships was accompanied by a tremendous expansion of world trade. By World War II, the age of commercial wind-propelled ships had ended, but during that war, ship-building, marine transportation, and naval operations rose to heights never before seen in history. Between 1942 and 1949, American shipyards alone produced nearly 5,000 steel merchant ships of over 2,000 gross tons each. These shipyards simultaneously constructed approximately 111,000 Navy combatant and auxiliary vessels including submarines, not to count vessels built for other defense departments. Since then, the significance of ships has held steady. Today more than 95% of all international cargo is transported by sea, and our reliance on defensive ships and submarines is as strong as ever.

For most of history, ships increased in efficiency primarily through simple trial and error. However, the thrust toward greater

size, speed, and efficiency has magnified the need for more sophisticated ship planning and design. Shipyards cannot afford to apply
trial and error to the construction of gigantic ships and submarines
costing billions of dollars and requiring many years to design and
build.

The upsurge in ship size and speed has also magnified the
difficulties associated with designing ships. Consider a modern
ship plying the high seas, attended by a complex, unsteady flow
field. The moving ship generates turbulence, spawning foaming
waves that trail into its wake and are joined by vortices twisting into
the depths like miniature tornadoes. Fields of bubbles churn from
the spilling waves and the spinning propeller.

All of these processes speak of forces that resist the hull's forward motion and influence its efficiency, speed, and maneuverabil-
ity. Resistance, created by pressure and friction forces against the
hull, is affected by waves generated by the hull, the hull's wetted
surface area, its shape and any appendages (including the rudder),
and speed. Resistance is increased by turbulence, produced largely
by flow dragging along the hull's surface. This layer of flow dragging along the hull is termed the boundary layer, and it increases in
thickness as the ship's length increases. If these interactions are not
complex enough, add the variable state of the sea—the ever-changing wind, waves, and currents—and the pitching and other motion
of the ship. These complexities further compound the resistance the
ship encounters and translate into its potential speed and efficiency,
the amount of energy required to propel it forward.

Other problems to consider include the rumble of the ship's propeller and the vibrations the propeller can produce in the ship's
hull. Noise, especially important for the stealth of defense ships, can
also be generated by the explosion-like collapse of bubbles formed
by propellers rotating so rapidly that they vaporize some of the
water passing between their blades, a process called cavitation.
These collapsing bubbles can cause serious damage to the propeller
and ship's hull. Bubbles from various sources leave behind a dis-
tinctive signature wake that provides an easy mark for enemy torpedoes. A thorough understanding of a ship's signature is essential
if it is to be minimized or if defensive counter-measures are to be
developed.

All of these factors must be considered in the design of modern ships, although their importance differs with the ship's function. Defense ships such as aircraft carriers, destroyers, frigates, and submarines need to be fast, maneuverable, stable, and quiet. On the other hand, efficiency—minimizing the amount of fuel required—is of primary importance for commercial tankers and container ships, as well as transport vessels.

Fortuitously, the development of modern vessels has been accompanied by the creation of more capable tools for modeling and analyzing ship performance: first towing tanks, experimental facilities and instruments, and concepts of model similitude, and then ever-more-powerful computers. Through the years, increasingly sophisticated tools and experimental techniques have been applied to the same complex flow questions, yielding ever-more-accurate results and predictions. IIHR has been involved in the evolution of these hydraulic and numerical-modeling tools since around the middle of the century because of a series of defense-related projects funded from the start by the U.S. Navy.

This relationship with the Navy, which has transformed IIHR into a leading ship-hydrodynamics research program far from any major body of water, commenced during World War II. It was initiated by Anton Kalinske, a creative institute researcher who had become active in fundamental turbulence and sediment-transport studies as well as more applied efforts related to plumbing. The Navy had been testing scale models of new ship designs in its Experimental Model Basin since around the turn of the century. In 1939, a modernized Navy research center, the David Taylor Model Basin (DTMB, later the Naval Surface Warfare Center, Carderock Division) first opened its doors. Kalinske established contact with DTMB at the beginning of the war and, according to Rouse, he "brought several of its projects to Iowa, for further study." Among these were investigations of the drag on stationary ships in flowing water, which were performed at IIHR on a small scale to determine the problems that might be encountered at DTMB in a similar but larger setup. These tests marked IIHR's initial investigations in ship hydrodynamics.

Kalinske also instigated the construction of the institute's first water tunnel in 1943, which was immediately put to use performing cavitation studies of projectiles. These studies, which were initiated by the wartime National Defense Research Committee (NDRC), were at the war's end taken over by the Navy's Bureau of Ships. Their continuation was funded by a broad contract that allowed fundamental questions to be investigated as the need arose. Navy-funded turbulence, cavitation, sediment-transport, and pressure-distribution studies supplanted ship hydrodynamics for the coming decade.

When Kalinske accepted an industrial position in 1945, his efforts were adopted by John McNown from the Navy's Radio and Sound Laboratory. A few years later, in 1948, the Bureau of Ships contract was transferred to the recently formed ONR, which expanded the Navy's support of the institute to include, for example, construction of new wind and water tunnels.

Thus, by the middle of the century, ONR had become a dominant and constant player in the institute's research agenda. ONR has remained a major funder through the twentieth century, although its broad post-WWII coverage of basic research has narrowed considerably. DTMB also has continued to support institute research and was a major contributor to IIHR in another way: through providing Lou Landweber, a senior staff member of 22 years at DTMB and head of its Hydrodynamics Division, who came to IIHR in 1954.

During and after World War II, Hunter Rouse visited DTMB regularly to consult about cavitation studies that were being funded by DTMB. Landweber and Rouse had become good friends during these visits. When McNown accepted a position at the University of Michigan, Rouse enticed Landweber to fill the vacated spot by offering Landweber an immediate full professorship. Landweber, a theoretician and experimentalist with training in mathematics and physics, later explained that professorial life was preferable to a job in which priorities were set by the military, and that Rouse's offer was thus impossible to refuse.

Landweber arrived simultaneously with an agreement from ONR to fund the conversion of the institute's basement river channel, built in 1919 as part of the original Hydraulics Laboratory, into a ship-towing tank—a narrow channel through which small-scale ship models would be pulled and assessed under various circumstances. This tank, then one of the few in the nation, solidified IIHR's capabilities to experimentally examine the motion of a ship. Landweber immediately took over theoretical and experimental

Lou Landweber reinitiated IIHR's ship-hydrodynamics studies in the mid-1950s and remained active in their execution for nearly 40 years.

studies connected to naval architecture and ship hydrodynamics. Within a year, Rouse was expressing relief that Landweber had also aptly assumed most administrative responsibility for naval projects.

Through the following decades, Landweber was instrumental in maintaining a leadership role in ship hydrodynamics here at Iowa. Even after his retirement in 1982, he remained an extremely active emeritus professor, receiving grants, publishing papers, and mentoring graduate students until 1995, when health problems restricted him to his home. He continued to carry out basic research in questions of wave and viscous resistance of ships, vibration, and in later years the Lagally theorem of ideal-fluid theory. Occasionally he ventured into applied research, in 1969, for example, testing the oil company ESSO's new supertanker. However, nearly all of his research continued to be funded by the U.S. Navy, from which he consistently received anywhere between two and a dozen grants, through ONR and the Naval Surface Warfare Center (formerly DTMB). Landweber continued experimental towing-tank studies to investigate ship resistance. He made major theoretical contributions to the prediction of waves around ship hulls, as well as to the prediction of resistance due to waves (using ideal-fluid theory) and of friction (using boundary-layer theory). A distinguished and widely recognized leader in these endeavors and theoretician whose insights extended well beyond the ordinary, he completed his studies primarily with his keen mathematical senses and his trusty slide rule.

The arrival of Virendra C. Patel in 1971 expanded IIHR's investigations of real (viscous) flows and initiated numerical modeling of these flows here. Patel was a leader in attempts to develop numerical solutions to complex, three-dimensional flows, and his efforts had caught the attention of Landweber and others in ship hydrodynamics. He focused, in particular, on viscous flows and turbulent boundary layers around ship hulls and other bodies immersed in fluids. His numerical applications were compared to experimental results, which could be more easily obtained in a wind tunnel than in a ship-towing tank. Patel's efforts spearheaded the incorporation of turbulent flows into ship-hydrodynamics efforts and led in 1982 to IIHR becoming an ONR Special Focus Research Center. Prior to this four-year award, the dominance of ship hydrodynamics relative to other IIHR initiatives had shriveled both in project number and total funding. The prestigious Special Focus grant immediately boosted the ship-hydrodynamics budget from $40,000 in 1981 to $112,000 in 1982 and to $251,000 the following year.

When the Special Focus grant was received, Patel resigned as chair of the college's Division of Energy Engineering so that he and Landweber could coordinate the efforts of a group of researchers, for the grant included funding for new faculty as well as student support. Together with post-doctoral associate Hamn-Ching Chen and others, the expanded research group was charged with using modeling to improve understanding of the flow of water around the entire ship, including considerations of wave generation, friction, and turbulence.

Through this project, major advances were made in both experimental and computational modeling techniques. Computational fluid dynamics (CFD) became established as a major institute endeavor, proving its worth through the project's development of RANSTERN, the first successful computer code for calculating viscous turbulent flow of water around submerged ship sterns and wakes, which became critical in stimulating subsequent CFD efforts here and elsewhere (see chapter 13, page 192).

The Special Focus grant allowed the 1983 hiring of a new faculty member, naval architect Fred Stern. Stern, who had been working in the three-dimensional numerical modeling of viscous flows, came to Iowa with two intents: introducing free-surface effects into the viscous-flow models and working on propeller–hull interactions. Stern's success in these two areas has resulted in his assuming the leadership of IIHR's ship-hydrodynamics efforts. He continues to bring in substantial and prestigious long-term grants, with funding increasing until in 1997, his research group received $600,000 from ONR—an amount that constituted the largest core funding for basic research allotted to any ONR fundee. This supports the vast majority of IIHR ship research. Stern heads an active research group, in 1997 consisting of research scientist Eric Paterson and postdoctoral associates Madhu Sreedhar, Joe Longo, and Bob Wilson, along with graduate students. The research group, along with its predecessors, has thus maintained IIHR's leadership role in ship-hydrodynamics research, managing to merge theoretical and applied studies looking at ideal and viscous fluids, and using experimental and computational efforts to help develop a practical understanding of flow from bow to wake. IIHR has also promoted ship hydrodynamics by providing crucial training to graduate students who have then gone on to work in the Navy and elsewhere on related computational efforts.

Much of Stern's work has focused on free-surface ship hydrodynamics, that is, experimentation and modeling of ships moving through the deforming air–water interface. Here turbulence, waves, and bubbly wakes magnify the complexity of the ship's movement and compound the complexity of experimental and numerical efforts. Yet, as described earlier, reducing the resulting resistance and minimizing the ship's hydrodynamic signature is crucial for its efficient operation and protection. Prior to Stern's efforts, IIHR's modeling dealt only with bodies that were completely surrounded by fluid or ignored the effects of the free surface. "Stern took the crucial step by incorporating the turbulent free surface," states Edwin

IIHR is one of the few laboratories that maintains personnel and equipment to perform both sophisticated physical and numerical ship modeling. The two interplay, with each supporting the accuracy and validity of the other. Here the flow around a U.S. Navy guided missile destroyer (top) is modeled at IIHR both physically (center) and computationally (bottom).

Rood, an ONR Program Officer, "and this marked a turning point in naval hydrodynamics."

Stern's efforts with free-surface flows have for several years emphasized detailed modeling, applied to specific situations or problems, coupled with experimental work. These took a new direction that marked another turning point for both the Navy and IIHR in 1994. That year, at the CFD Workshop in Tokyo, IIHR's codes were judged to be among the best for the development of ship-hydrodynamics codes. In addition, the extensive data on free-surface interactions collected with IIHR's towing tank were crucial in validating computational models at this workshop. The new opportunities presented by CFD, at a time when ONR research was mandated to be increasingly useful to the Navy, convinced the Navy to encourage the transition of CFD codes to actual ship design. The desired codes would need to be fast, accurate, practical, and generally applicable. Their development would be costly. Thus, the Navy downsized to support only two research groups, one of which was IIHR's. This selection expanded IIHR's ONR funding tremendously, but it also added new stipulations. Stern and his group, supplemented by the hiring of Paterson as a research scientist, now added the development of a general-purpose code to their other research goals. ONR's funding is accompanied by the stipulation that IIHR's research team must be able to transfer its computer code for use by Navy engineers. The insistence on quality control for both computational and experimental work has also been magnified. That emphasis has led to Stern's development, in collaboration with Hugh Coleman, of a new theory for quantitatively evaluating numerical and modeling uncertainty.

Advances in the code are being closely tracked and validated by experiments performed in the towing tank, which has been completely upgraded since 1993. The old wavemaker has been replaced by a new one with a computer-controlled plunger. With the aid of an independent wave-measuring system electronically linked to the drive carriage, a moving ship's exact relationship to a series of pulsing waves now can be recorded at every point along the towing tank. A unique particle image velocimeter (PIV), in 1997 the only moving underwater PIV in the world, has also been installed to measure components of velocity and turbulence as a model ship hull is towed through a simulated rough ocean. This towing-tank

The success of IIHR's ship-hydrodynamics studies depends on teamwork combined with the fruitful interplay of experimental and numerical modeling efforts. Here Fred Stern (far left) and his 1992 research team, (from left to right) Joe Longo, Eric Paterson, Jung-Eun Choi, and Yusuke Tahara, discuss the setup of instrumentation that will be used to collect data on velocities and pressures in the water surrounding the model ship hull (background) as it is being towed. The mean pressures and velocities will then be used to validate CFD codes that will numerically model water movement around the hull.

equipment forms a precise and integrated experimental unit for ac-curately describing the interactions between a ship's hull and the water through which it glides. Towing-tank experiments, employ-ing both traditional and advanced equipment, use both idealized and practical geometries for physical understanding and to support CFD code development and validation. Here again, there is great emphasis on experimental uncertainty analysis.

This combined expertise in experimental and computational modeling is an example of one of IIHR's major strengths. While many ship-hydrodynamics laboratories specialize in either one or the other type of modeling, IIHR is one of the few academic labora-tories anywhere that performs both on a large scale and with a high degree of sophistication, and that also concerns itself with fundamental questions. With the resulting interplay of concepts and data, Stern says that soon his group will succeed in using CFD codes to design experiments in the towing tank.

Stern's research has also investigated interactions between the gigantic rotating propellers, which can be 20 feet or even larger in diameter, and a ship's hull. Together the hull and propeller form an interactive system, each influencing the flow field and thereby the performance of the other. In an effort to minimize the interactions with the ship's hull, blades are becoming increasingly more complicated. Stern's research team is undertaking a project to decipher the great complexities of water's flow around propellers and to develop CFD codes that simulate real flows between propeller blades. Efforts are focused on development of CFD methods for more advanced propulsion concepts.

These multiple IIHR efforts, combined with those of collaborators located elsewhere, are contributing to the creation of an integrated system that attempts to consider all components of ship hydrodynamics. As Edwin Rood in 1997 summarized, "Today IIHR, in the person of Fred Stern, leads an international effort to develop unsteady RANS surface ship-hydrodynamics codes using complementary experimental/numerical procedures for verification and validation. There is no other complex engineering problem that I know of which is being attacked so comprehensively and so successfully." When the understanding of the mathematics and programming of complex flows is precise and accurate enough to be predictive, the new field of simulation-based design will emerge in full force and revolutionize the ship- and submarine-design process. Stern can see his research moving toward that end. In the 1980s, resistance could be approximated only through models of ship hulls towed along a towing tank. Those results were then extrapolated to assess resistance of a full-scale ship. In 1997, supercomputers could predict the resistance of a real ship with 95% accuracy. In time, this accuracy will be improved even further. "Today we have the Model T of simulated design," Stern states. "It's crude and basic, but it works." With his associates, he is striving for the day when, with the press of computer keys, the forces placed on a maneuvering ship by the heaving and churning oceans become virtual reality on a computer screen, and a ship's structure—before any steel is cast—can be optimally molded to suit its purpose.

20. Fundamental Fluid Mechanics: Vortex Dynamics

If it weren't for turbulence and its chaotic cousins, vortices, IIHR would be out of business. Without turbulence, our existing computer technologies would be adequate to solve fully the equations of fluid flow and we'd have relatively simple solutions to most of our questions.

Such simplicity is not to be found. Most flows in nature are turbulent, and vortices are ubiquitous. The sheaths of air and water that surround the earth are constantly swirling, with eddies ranging in scale from a few millimeters to the size of a continent. The ocean, viewed from a satellite, is composed of a churning mass of gigantic vortices. Rivers contain minuscule vortices as well as vortices that may occupy entire backwater regions behind structures such as wingdams. Vortices spin off the wingtips of bumblebees and birds, enabling highly efficient flight. More dramatic atmospheric vortices form devastating cyclones, hurricanes, and tornadoes. Dust devils twist upward from bared soil in school playgrounds. Currents caused by the clash of incoming and outgoing tides breed tidal whirlpools. Warm-air vortices generated by the natural convective rising of heated air are utilized by soaring birds, who choose their position so they will be lifted upward. Looking out on dark nights, we might see the spiral of the Whirlpool Galaxy, or imagine the rotating swirl of infalling matter into black holes. Students of vortex dynamics will never lack subject matter. Nor will they lack motivation, for vortices and turbulence are equally as numerous in and around human constructs such as bridge piers, where they confound all levels and types of questions that hydraulic engineers are trying to answer. Indeed, vortices are key components of innumerable flow phenomena, and they pose some of the most challenging problems in fluid physics.

The need for reliable models of complex turbulent flows underpins most fluids-engineering efforts, from those addressing river and coastal sediment erosion to projects involving turbomachinery operation or the turbulent boundary layer around ships. Research at IIHR, although often applied to specific problems and situations, has always also incorporated studies that attempt to shed light on the fundamentals of complex fluid flow. These efforts have been aimed at examining the underlying physics of turbulence, with the hope of ultimately developing new, improved models for turbulence and techniques for measuring and simulating turbulent flow. Researchers involved in such efforts historically have attempted to model turbulent flows through a comparison of experimental observation and numerical simulation, in which the effects of turbulence are accounted for by adding new terms, suggested by experiments, to the equations traditionally used to determine momentum and energy in fluids. Most of IIHR's research projects, regardless of ultimate purpose, must incorporate some model for the turbulence that is ubiquitous and often problematic.

An example of IIHR's multiple, ongoing efforts with turbulence is the work of Jeff Marshall, one of IIHR's experts on turbulence and vortices. An examination of his work sheds light on the types of fundamental investigations that other IIHR researchers have undertaken through the years. Marshall is attempting to further our ability to model turbulence by studying the microphysics of the spaghetti of vortices that generate the turbulent fluctuations. By developing insight into the smallest ingredients of turbulence—for example, the generation and evolution of thin strands of vorticity, their interaction with a wall or with other vortices, or their interplay with solid particles—Marshall hopes to understand the basic ingredients underlying the chaotic dynamics of turbulent flow. Through dissecting the fundamental interactions of vortices in turbulent flows, he hopes to develop new types of turbulence models that are based directly on the dynamics of these micro-structural ingredients.

Some of Marshall's efforts have also involved development of new numerical algorithms for solution of fluid flow based directly on vorticity transport. Why is this important? Traditional CFD,

Doctoral student Veera Rajendran is studying fundamental properties of subsurface vortices in a model pump bay. Strong subsurface vortices in a pump adversely affect pump performance, possibly causing vibration damage and shortening the pump's life—a problem that lends relevance to this type of fundamental study. Rajendran uses a PIV system to obtain quantitative data on the strength, core size, and fluctuations of subsurface vortices attached to the floor, backwall, and sidewalls of the intake bay.

which is based upon direct solution of the Navier-Stokes equations for velocity and pressure, works well for tracing near-body flows such as those of boundary layers and within pipes. However, objects passing through a fluid shed vorticity into wakes that extend well beyond the point of generation. There, traditional CFD experiences numerical dissipation: the computer simulation spreads and loses the thin filaments of vorticity artificially because the simulation grid no longer can resolve the vorticity structures.

Marshall's vorticity methods instead solve directly for the vorticity field from the vorticity-velocity formulation of the Navier-Stokes equations. The flow solution is obtained on points that follow the vorticity; the method eliminates the fixed grid and instead relies on control points that travel along with the flow, integrating over moving and irregular grid spaces. This moving, dynamic numerical grid has major advantages. For one thing, the grid need only cover the part of the flow with significant vorticity (e.g., boundary layers or vortices shed in wakes). These flows typically make up a small fraction of the total flow field, and thus far fewer numerical solutions are needed with Marshall's techniques. In addition, because the control points move along with the flow, vorticity is never lost through artificial dissipation.

Much of Marshall's work on vortex methods focuses on extending these techniques from their traditional applications (to inviscid, two-dimensional flows) to viscous, three-dimensional flows. Marshall has even extended his vortex method to two-phase flows, that is, a flow with solid particles in a liquid or a gas, gas bubbles in a liquid, or liquid droplets in a gas. Two-phase flows are abundant in nature: rivers carry particles of sediment, bubbles rise in boiling water, raindrops fall through the air. Marshall observes these types of flows, asking how the existence of the second phase affects the flow's turbulence. He investigates how the vortices shed by the particles (or bubbles or droplets) interact with the turbulent vortices already present in the surrounding fluid, computing both how the fluid induces motion of the particles and how the fluid flow is affected by the particles.

While much of his work centers on developing computational or theoretical models for the evolution of vorticity, Marshall resembles other IIHR researchers in also being an experimentalist. He uses laboratory studies to validate computational models and to suggest new models. Marshall might, for example, use a laser light sheet to visualize regions of vorticity-laden flow. The light sheet excites fluorescent dye injected in the water, producing an image that is recorded by a video or still camera. Quantitative measurements of the vorticity field are obtained using a second technique called particle image velocimetry (PIV). PIV utilizes a pulsed laser to produce a sheet of light that reflects off seed particles that are moving with the swirling water. The reflections of two closely spaced light pulses are detected on a single photographic or video image, which Marshall then analyzes for the speed and direction of motion of each particle. Both these techniques are ideal for studying the intricacies of the dynamics of thin vortex structures because, unlike the pitot tubes used in the past, these newer techniques are nonintrusive and do

Jeff Marshall (standing) discusses with student Meihong Sun results obtained from her use of a PIV system to study vortex interaction with the wake of a sphere. The PIV system uses the Nd:YAG laser (seen in the rear) and an optical guidance system to form a sheet of light emitted into the vortex generator tank (not shown). The equipment directly to the left of the workstation synchronizes the flashing of the laser with photographic-image acquisition. The photographic images are digitized and fed into the workstation, where they are processed and displayed on the screen as a plane of velocity vectors.

not disrupt the water's flow. PIV is essential for measurement of vorticity in unsteady flows because it provides velocity data instantaneously over the entire two-dimensional plane illuminated by the laser, whereas earlier techniques (such as hot-wire anemometry or laser Doppler velocimetry) can measure velocity only at a single point.

Although IIHR's initial PIV setup was purchased for fundamental turbulence investigations, use of this sophisticated technology has since spread to other types of projects. For example, Allen Bradley, Anton Kruger, and Marian Muste have applied PIV techniques to their analysis of river flows. Fred Stern's research group is employing a unique underwater PIV system in the towing tank, where it is contributing to ship-hydrodynamics studies. This sharing of state-of-the-art technologies among researchers and projects is not uncommon at IIHR, where the rich mixture of applied and fundamental projects, and of researchers with a variety of interests, profits everyone.

Like much IIHR fundamental research, Marshall's work also has more immediate relevance to a wide range of practical applications. His research on vortex-body interaction in viscous fluid, for example, is important for predicting helicopter vibrations caused by vortices spinning off the blade tips of the main rotor and crashing into the vehicle tail section. His research on turbulence interaction with strong vortex structures and on three-dimensional vortex instabilities is important for modeling decay of aircraft trailing vortices, a vital issue for airport safety.

Vortex dynamics and computational and experimental techniques that concentrate on the vorticity field are studied in numerous ongoing IIHR projects undertaken by other researchers. Vortices, for example, spin off the bows of moving ships and their propellers, controlling the bubble distributions in wakes that motivate many of Stern's ship-hydrodynamics studies. The air entrained in vortices generated within massive dropshafts stimulate Jain's studies of these hydraulic structures. Vortices commonly form in pump intake structures that draw in cooling water for power plants. There they reduce the efficiency of the pumps and may entrain sediment or air bubbles that damage turbine blades, problems that motivate Nakato's water-intake investigations and Patel's efforts to numerically model these flows. An abundance of applied-research projects through the years, performed by numerous IIHR research scientists, have involved vorticity and two-phase flow, such as the scour of soil from the base of bridge piers, transport of sediment in rivers and around hydraulic structures, and movement of raindrops through the air. In addition, fundamental elements of vorticity have been examined, for example, by Patel through his research on vortex generation by three-dimensional boundary-layer separation, and by Ching-Long Lin through his attempts to simulate large eddies of the planetary boundary-layer flows.

Through the years, research at IIHR has fully spanned the continuum from totally fundamental to site specific, applied studies. As a collective unit, the research has posed and answered questions that range from the physics of fluid movement to how to solve very practical problems. Research into the generation and effects of vorticity is no different. Although Marshall's work would be placed at one end of that continuum, his studies meld with numerous others to present a unified and comprehensive attack on the enigmas of turbulence and vortex-ridden flows that will continue to dominate hydraulics studies for the indefinite future.

Appendix A.
IIHR Senior Staff, 1971–97

Alonso, Carlos; Research Engineer; 1971

Ames, William F.; Professor and Research Engineer; 1967–75

Below, Paul; Senior Accountant; 1994–present

Bradley, A. Allen, Jr.; Assistant Professor and Research Engineer; 1994–present

Bravo, Hector; Postdoctoral Associate; 1990–91

Carpenter, Theresa; Research Associate; 1993–94

Chandrasekhara, M. S.; Postdoctoral Associate; 1983

Chang, Kaung Jain; Postdoctoral Associate; 1975

Chen, Ching-Jen; Professor, Chairman ME and Research Engineer; 1971–92

Choi, Do-Hyung; Postdoctoral Associate; 1978–79

Chwang, Allen T.; Professor and Research Engineer; 1978–92

Cramer, James; Data Systems Coordinator; 1981–92

Croley, Thomas E., II; Associate Professor and Research Engineer; 1972–80

Dague, Richard; Professor and Research Engineer; 1975–78

Danushkodi, Vaiyapuri; Postdoctoral Associate; 1975

DeJong, Darian; Engineer I; 1996–present

Denli, Nuray; Postdoctoral Associate; 1978–79

Diplas, Panayiotis; Postdoctoral Associate; 1986–88

Dlubac, James J.; Postdoctoral Associate; 1983

Eichinger, William; Associate Professor and Research Engineer; 1997–present

Note: For each person, only the last position held at IIHR is indicated; however, inclusive years of IIHR employment are shown. The word "present" indicates employment continuing into 1998. Staff members prior to 1971 are shown in a chart appended to this listing. The chart has been reproduced from *The First Half Century of Hydraulic Research at the University of Iowa*, IIHR Bulletin #44.

Ettema, Robert; Professor and Research Engineer; 1980–present

Farell, Cesar; Associate Professor and Research Engineer; 1971–79

Fischer, Edward E.; Research Associate; 1979–80

Georgakakos, Konstantine P.; Associate Professor and Research Engineer; 1985–94

Giaquinta, Arthur R.; Adjunct Associate Professor and Associate Research Scientist; 1974–86

Glover, John R.; Professor and Research Engineer; 1969–80

Goss, James; Engineer III; 1978–present

Güven, Oktay; Postdoctoral Associate; 1976

Harris, Dale; Assistant Director for Operations; 1940–84

Hartman, Kenneth M.; Engineer I; 1983–86

Holly, Forrest M., Jr.; Professor, Chairman CEE, and Research Engineer; 1982–present

Houser, Douglas; Engineer II; 1986–present

Howe, Joseph; Professor Emeritus; 1929–83

Huang, Ching-Jer; Postdoctoral Associate; 1991–92

Jain, Subhash C.; Professor and Research Engineer; 1970–present

Janssen, Marlene; Administrative Associate; 1975–present

Johnson, J. Kent; Assistant Research Scientist; 1983–present

Johnson, Timothy; Postdoctoral Associate; 1997

Jovic, Srboljub D.; Postdoctoral Associate; 1985–87

Karim, Fazle; Assistant Research Scientist; 1981–86

Kennedy, John F.; Director, Research Engineer, and Hunter Rouse Professor of Hydraulics; 1966–91

Kim, Hyoung-Tae; Postdoctoral Associate; 1989–92

Kim, Sungeun; Postdoctoral Associate; 1991

Kim, Wu Joan; Postdoctoral Associate; 1992–94

Kitanidis, Peter; Assistant Professor and Research Engineer; 1979–84

Krajewski, Witold F.; Associate Professor and Research Engineer; 1987–present

Kruger, Anton; Assistant Research Scientist and Adjunct Assistant Professor; 1992–present

Landweber, Louis; Professor Emeritus; 1954–98

Lekakis, Ioannis C.; Assistant Professor and Research Engineer; 1987–88

Li, Peter; Postdoctoral Associate; 1978–79

Li, Yitian; Postdoctoral Associate; 1990–91

Lin, Ching-Long; Assistant Professor and Research Engineer; 1997–present

Lin, Jung-Tai; Research Engineer; 1969–70

Locher, Frederick A.; Research Engineer; 1967–74

Longo, Joseph F.; Postdoctoral Associate; 1996–present

Macagno, Enzo O.; Professor Emeritus; 1956–present

Macagno, Matilde; Assistant Professor and Assistant Research Scientist Emeritus; 1959–present

Marshall, Jeffrey; Associate Professor and Research Engineer; 1993–present

Matel, Mark; Systems Programmer I; 1995–97

McDonald, Donald B.; Professor Emeritus; 1973–present

McDougall, David; Research Engineer; 1968–71

Melville, Joel G.; Research Engineer and Adjunct Assistant Professor; 1972–76

Menendez, Angel; Postdoctoral Associate; 1983–84

Meselhe, Ehab; Postdoctoral Associate and Adjunct Assistant Professor; 1994–97

Muste, Marian; Postdoctoral Associate; 1995–present

Mutel, Cornelia F.; Program Associate; 1990–present

Nakato, Tatsuaki; Adjunct Professor, Associate Director, and Research Scientist; 1975–present

Nakayama, Akihiko; Postdoctoral Associate; 1975

Nixon, Wilfrid A.; Associate Professor and Research Engineer; 1987–present

Noblesse, Francis; Postdoctoral Associate; 1975

Odgaard, A. Jacob; Professor, Research Engineer, and Associate Dean CoE; 1977–present

Ogden, Fred; Adjunct Assistant Professor and Postdoctoral Associate; 1992–95

Onishi, Yasuo; Research Engineer; 1972–74

Paily, Poothrikka P.; Assistant Research Scientist; 1974–75

Parrish, John; Postdoctoral Associate; 1995–97

Parthasarathy, N. Ramkumar; Assistant Research Scientist; 1989–93

Patel, V. C.; Director, Research Engineer, and University of Iowa Foundation Distinguished Professor; 1971–present

Paterson, Eric; Assistant Research Scientist and Adjunct Assistant Professor; 1994–present

Raja Rao, Kuchibhotla; Postdoctoral Research Fellow; 1976–77

Ramaprian, Belakavadi R.; Professor and Research Engineer; 1976–85

Rexroth, David; Research Assistant; 1990–91

Rodriguez-Iturbe, Ignacio; Endowed Professor of Civil Engineering; 1989–91

Rouse, Hunter; Carver Professor Emeritus; 1939–76

Sancholuz, Arturo; Assistant Research Scientist; 1975–76

Sato, Chikashi; Postdoctoral Associate; 1981–85

Savic, Ljubodrag; Postdoctoral Associate; 1992–93

Sayre, William W.; Professor and Research Engineer; 1968–80

Schiller, Eric; Assistant Research Scientist and Adjunct Assistant Professor; 1973–74

Schnoor, Jerald; Professor and Research Engineer; 1978–present

Schwarz, Joachim; Associate Research Scientist and Adjunct Assistant Professor; 1971–74

Shen, Huihua; Postdoctoral Associate; 1991–92

Singerman, Robert; Assistant Research Scientist; 1974–75

Sinha, Sanjiv; Postdoctoral Associate and Adjunct Assistant Professor; 1996–97

Sirviente, Ana; Postdoctoral Associate; 1996–97

Sotiropoulos, Fotis; Assistant Research Scientist; 1991–95

Sreedhar, Madhu; Postdoctoral Associate; 1994–98

Stern, Frederick; Professor and Research Engineer; 1983–present

Tahara, Yusuke; Postdoctoral Associate; 1992–94

Tang, Chii-Jau; Postdoctoral Associate; 1987

Tatinclaux, Jean-Claude; Associate Professor and Research Engineer; 1969, 1973–82

Tracy, Kenneth; Assistant Professor and Research Engineer; 1974–75

Tsintikidis, Dimitris; Research Associate; 1994–95

Wang, Yalin; Postdoctoral Associate; 1991–93

Weber, Larry; Assistant Professor and Research Engineer; 1994–present

Wei, Yingchang; Postdoctoral Associate; 1995–98

Weirich, Frank; Associate Professor and Research Engineer; 1993–present

Whelan, Andrew; Postdoctoral Associate; 1996

Wilson, Mark A.; Data Systems Coordinator; 1989–present

Wilson, Robert; Postdoctoral Associate; 1997–present

Wu, Han-Chin; Associate Professor and Research Engineer; 1972–76

Yoon, Byungman; Postdoctoral Associate; 1991

Yoon, Joon-Yong; Postdoctoral Associate; 1993–94

Zhang, Daohua; Postdoctoral Associate; 1991–93

Zhang, You-Kuan; Professor and Research Engineer; 1994–present

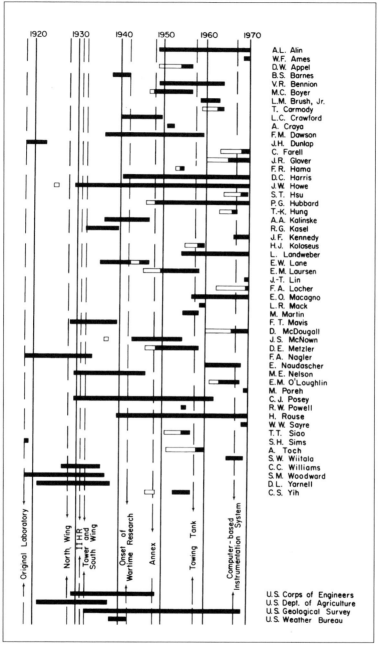

IIHR personnel, 1920–70

Appendix B.
Graduate Degree
Recipients, 1971–97

Ahn, Kyungmo. Evaluation of the Accuracy of a New Settling-Tube Method for Grain-Size Analysis. M.S., 12/86. Advisors: T. Nakato and J. F. Kennedy.

Aksoy, Hakan. Finite Analytic Numerical Solution of Fluid Flow and Heat Transfer with Non-Staggered Grids. M.S., 5/89. Advisor: C.-J. Chen.

Akyeampong, Yaw Appiah-Nuamah. The Local Sediment Transport Rate in Periodically Non-Uniform Flows. Ph.D., 1/71. Advisor: J. F. Kennedy.

Akyurek, Murat. Sediments Suspension by Wave Action over a Horizontal Bed. M.S., 7/72. Advisor: F. Locher.

Al-Gamdhi, Abdullah. No thesis. M.S., 12/87. Advisor: S. C. Jain.

Albert, Dale T. No thesis. M.S., 7/87. Advisor: C.-J. Chen.

Ali, Matahel Ag Mohamed. Hydraulic Characteristics of Helicoidal-Ramp Inlets for Vortex Flow Dropstructures. M.S., 8/93. Advisor: S. C. Jain.

Allen, Mark Edward. An Analytical Model for Flow through Vertical Barrier Screens. M.S., 12/96. Advisor: L. Weber.

Alonso, Carlos V. Time-Dependent Confined Rotating Flow of an Incompressible Viscous Fluid. Ph.D., 1/71. Advisor: E. O. Macagno.

Anagnostou, Emmanouil Nikolaos. Three-Dimensional Simulation of Radar-Rainfall Observations. M.S., 8/94. Advisor: W. F. Krajewski.

Anagnostou, Emmanouil Nikolaos. Real-Time Radar Rainfall Estimation. Ph.D., 8/97. Advisor: W. F. Krajewski.

Anderson, David Ray. Computer-Aided Analysis of Two-Dimensional, Transient Heat Conduction Problems. M.S., 8/87. Advisor: C.-J. Chen.

Note: Graduate theses completed prior to 1971 are listed in *The First Half Century of Hydraulic Research at the University of Iowa*, IIHR Bulletin #44.

Ansar, Matahel. Experimental and Theoretical Studies of Pump-Approach Flow Distributions at Water Intakes. Ph.D., 8/97. Advisors: T. Nakato and V. C. Patel.

Argue, John R. The Mixing Characteristics of Submerged Multiple-Port Diffusers for Heated Effluents in Open Channel Flow. M.S., 5/73. Advisor: W. W. Sayre.

Ashton, George D. The Formation of Ice Ripples on the Underside of River Ice Covers. Ph.D., 5/71. Advisor: J. F. Kennedy.

Baban, Farzad. Mean Flow Measurements in the Tail Region and the Near-Wake of a Body of Revolution at Incidence. M.S., 12/83. Advisor: B. R. Ramaprian.

Bae, Deg Hyo. A Comparative Study of Three Spatially-Lumped Soil Models for Real-Time Flood Prediction. M.S., 12/89. Advisor: K. P. Georgakakos.

Bae, Deg Hyo. Hydrologic Modeling for Flow Forecasting and Climate Studies in Large Drainage Basins. Ph.D., 8/92. Advisor: K. P. Georgakakos.

Baek, Je-Hyun. Three-Dimensional Turbulent Boundary Layers on the Bodies of Revolution at Incidence. Ph.D., 5/84. Advisor: V. C. Patel.

Baires, Manual. No thesis. M.S., 12/78. Advisor: E. O. Macagno.

Balasubramanian, Velaydhan. Horizontal Buoyant Jet in Quiescent Shallow Water. M.S., 7/76. Advisor: S. C. Jain.

Barkdoll, Brian David. Sediment Control at Lateral Diversions. Ph.D., 5/97. Advisors: R. Ettema and A. J. Odgaard.

Bauer, Deborah Isabel. Subsurface Vortex Suppression in Water Intakes with Multiple-Pump Sumps. M.S., 5/96. Advisors: T. Nakato and R. Ettema.

Bergs, Mary Agnes. Flow Processes in a Curved Alluvial Channel. Ph.D., 5/90. Advisor: A. J. Odgaard.

Bhattacharya, Protosh Kumar. Sediment Suspension in Shoaling Waves. Ph.D., 1/71. Advisor: J. F. Kennedy.

Bhiladvala, Rustom. No thesis. M.S., 5/86. Advisor: C.-J. Chen.

Black, Richard. No thesis. M.S., 5/91. Advisor: F. Stern.

Bravo, Hector Ramon. Flow Fields in Navigation Installations Induced by Hydropower Releases. Ph.D., 8/89. Advisors: S. C. Jain and F. M. Holly, Jr.

Bravo, Ramiro Humberto. Computer-Aided Analysis of Two-Dimensional Fluid Flow and Convective Heat Transfer. M.S., 12/87. Advisor: C.-J. Chen.

Brockway, Charles G. An Evaluation of Streamflow Reconstructions for Estimation of Drought Statistics. Ph.D., 12/95. Advisor: A. A. Bradley.

Brown, Michael L. Experimental Investigation of the Interaction between a Longitudinal Vortex and a Turbulent Boundary Layer. M.S., 5/85. Advisor: B. R. Ramaprian.

Brucker, Gregory C. No thesis. M.S., 12/90. Advisor: C.-J. Chen.

Bugler, Thomas William, III. Scale Effects on Cooling Tower Model Studies. M.S., 12/74. Advisor: J. C. Tatinclaux.

Calisto, Oswaldo. No thesis. M.S., 12/80. Advisor: P. K. Kitanidis.

Caradenas, Alejandro. Prevention of Thermal Wedges Resulting from Multiple Port Diffusers. M.S., 7/73. Advisor: W. W. Sayre.

Caro-Cordero, Rodrigo. Mixing of Power-Plant Heated Effluents with the Missouri River. M.S., 5/77. Advisor: W. W. Sayre.

Carpenter, Theresa M. GIS-Based Procedures in Support of Flash Flood Guidance. M.S., 8/93. Advisor: K. P. Georgakakos.

Carrasquel Vera, Saul C. Effect of Wind Tunnel Walls on the Flow about Circular Cylinders and Hyperbolic Cooling Tower Models. M.S., 12/74. Advisor: C. Farell.

Celik, Ismail. Mean Flow past Circular Cylinders. Ph.D., 12/80. Advisors: V. C. Patel and L. Landweber.

Chan, Daniel T.-L. Turbulent Nonbuoyant or Buoyant Jets Discharged into Flowing or Quiescent Fluids. Ph.D., 7/72. Advisor: J. F. Kennedy.

Chandrasekhara, Muguru Subramanyam. Study of Vertical Plane Turbulent Jets and Plumes. Ph.D., 5/83. Advisor: B. R. Ramaprian.

Chang, Kaung Jain. Compression of Columnar-Grained Ice and Some Further Aspects of Brittle Fracture. M.S., 12/74. Advisor: H. C. Wu.

Chang, King Yu. Investigation of Fluid Motion in a Recoil Mechanism. M.S., 7/78. Advisor: C.-J. Chen.

Chang, Kuo-Cheng. Calculation of Three-Dimensional Boundary Layers on Ship Forms. Ph.D., 5/75. Advisor: V. C. Patel.

Chang, Sen-Ming. Finite Analytic Numerical Solutions for Steady Two Dimensional Heat Transfer and Flow in Bends. M.S., 12/81. Advisor: C.-J. Chen.

Chang, Sen-Ming. Prediction of Turbulent Internal Recirculating Flows with k-e Models. Ph.D., 12/84. Advisor: C.-J. Chen.

Chang, Yung-Chi. Lateral Mixing in Meandering Channels. Ph.D., 5/71. Advisor: W. W. Sayre.

Chao, Ru-Shiow. No thesis. M.S., 12/79. Advisor: A. J. Odgaard.

Chen, Bih-Ling M. Daily Streamflow Time Series Analysis. M.S., 5/78. Advisor: T. E. Croley II.

Chen, Bin. Computational Fluid Dynamics of Four-Quadrant Marine-Propulsor Flow. M.S., 7/96. Advisor: F. Stern.

Chen, Ching-Hong. Deformation and Stress Analysis of the Left Ventricle—An Active Thin Wall Model. M.S., 5/79. Advisor: C.-J. Chen.

Chen, Hamn-Ching. Development of Finite Analytic Method for Unsteady Three-Dimensional Navier-Stokes Equations. Ph.D., 7/82. Advisor: C.-J. Chen.

Chen, Wen-Chung. Finite Analytic Numerical Solution of Supersonic Flows in Two-Dimensional Non-Symmetric Channels. Ph.D., 12/84. Advisor: C.-J. Chen.

Chen, Ya Chi. Diffracted Acoustic Waves Due to a Soft Sphere near Flat Wall. M.S., 12/91. Advisor: A. T. Chwang.

Cheng, Mow-Soung. Analysis of Different Types of Dry-Wet Cooling Towers. Ph.D., 5/76. Advisor: V. C. Patel and T. E. Croley, II.

Cheng, Shui-Tuang. Compressive and Shear Strength of Fragmented Ice Covers. M.S., 7/77. Advisors: J. C. Tatinclaux and J. F. Kennedy.

Cheng, Wu-Sun. Finite Analytic Numerical Solution for Two Dimensional Incompressible Flows over an Arbitrary Body Shape. M.S., 12/82. Advisor: C.-J. Chen.

Cheng, Wu-Sun. Finite Analytical Numerical Solutions of Incompressible Flow past Inclined Axisymmetric Bodies. Ph.D., 12/86. Advisor: C.-J. Chen.

Cherian, Mookencheril P. An Analytical Solution for Buoyancy-Driven Flow in Sidearms of Cooling Lakes. Ph.D., 5/85. Advisor: S. C. Jain.

Chiang, P.-T. No thesis. M.S., 5/73.

Chieh, Shih-Huang. A Model Study of Pump-Intake Vortices at the Lake Chicot Pumping Station. M.S., 12/77. Advisors: T. Nakato and C. Farell.

Chiou, Jenq-shing. Turbulent Heat Transfer to Liquid Metals in a Circular Pipe. Ph.D., 5/80. Advisor: C.-J. Chen.

Choi, Do-Hyung. Three-Dimensional Boundary Layers on Bodies of Revolution at Incidence. Ph.D., 12/78. Advisor: V. C. Patel.

Choi, Jung-Eun. Role of Free-Surface Boundary Conditions and Nonlinearities in Wave/Boundary-Layer and Wake Interaction. Ph.D., 12/93. Advisor: F. Stern.

Choi, Seok Ki. Numerical Study of Laminar and Turbulent Flows past Two Dimensional and Axisymmetric Bodies. Ph.D., 5/89. Advisor: C.-J. Chen.

Chung, Cheng-Hua. Development of Cutting Edges for Ice Removal from Pavements. M.S., 5/92. Advisor: W. A. Nixon.

Chung, Kyung-Nam. Application of a Zonal Approach and Turbulence Models to Foils and Marine Propellers. Ph.D., 12/93. Advisor: F. Stern.

Ciach, Grzegorz Jan. Radar Rainfall Estimation as an Optimal Prediction Problem. Ph.D., 8/97. Advisor: W. F. Krajewski.

Collado, Jaime. Optimization in Water Resources. Ph.D., 12/84. Advisor: P. K. Kitanidis.

Constantinescu, Serban George. Numerical Simulation of Flow in Pump Bays Using Near-Wall Turbulence Models. Ph.D., 12/97. Advisor: V. C. Patel.

Cook, Alfred Glen. A Study of Urea Ice. M.S., 5/83. Advisor: R. Ettema.

Copeland, Ronald Robert. Numerical Modeling of Hydraulic Sorting and Armoring in Alluvial Rivers. Ph.D., 5/93. Advisor: F. M. Holly, Jr.

Craig, William O. Measurement of Wall-Shear Stress in Three Dimensional Flows. M.S., 7/82. Advisor: B. R. Ramaprian.

Damian, Radu Mircea. An Investigation for Possible Invariant Solutions of the Navier-Stokes Equations. M.S., 7/72. Advisor: W. F. Ames.

Danushkodi, Vaiyapuri. An Experimental Investigation of the Turbulent Structure of Sediment Suspensions. Ph.D., 5/75. Advisors: C. Farell and A. Giaquinta.

Den Bleyker, Jeffrey Scott. Mitigation of Predation at a Juvenile Bypass Outfall Site. M.S., 5/96. Advisors: L. J. Weber and A. J. Odgaard.

Denli, Nuray. An Analytical Model of Flow Induced by Longitudinal Contractions in the Small Intestine. M.S., 12/75. Advisor: J. G. Melville.

Denli, Nuray. An Analytical Solution of the Thick Axisymmetric Turbulent Boundary Layer on a Long Cylinder of Constant Radius. Ph.D., 12/78. Advisor: L. Landweber.

Dinavahi, Surya Prasad G. Effect of Boundary Layer on Thrust Deduction. M.S., 7/81. Advisor: L. Landweber.

Diouf, Mohammed. Experimental Study of Unstable Interface in Porous Media. M.S., 12/79. Advisor: S. C. Jain.

Dlubac, James Joseph. Oscillatory Singularities in a Saturated Poroelastic Medium. Ph.D., 5/83. Advisor: A. T. Chwang.

Dolphin, Garth Warren. Evaluation of Computational Fluid Dynamics for a Flat Plate and Axisymmetric Body from Model- to Full-Scale Reynolds Numbers. M.S., 5/97. Advisor: F. Stern.

Eli, Robert Nelson. An Hourly Precipitation Model for Ralston Creek. Ph.D., 7/76. Advisor: T. E. Croley II.

Falcón Ascanio, Marco Antonio. Analysis of Flow in Alluvial Channel Bends. Ph.D., 5/79. Advisor: J. F. Kennedy.

Fang, Wen-Tsun. Uncertainty Analysis in Prediction of River Water Temperature. M.S., 5/91. Advisor: W. F. Krajewski.

Fang, Wen-Tsun. Sound Generation and Damping Effects by Neighboring Bubbles near a Free Surface. Ph.D., 12/94. Advisor: A. T. Chwang.

Fernández-Pichardo, César Augusto. Vortex Suppression by Means of Grids. M.S., 12/79. Advisor: E. O. Macagno.

Fischer, Edward Ernst. Scour around Bridge Piers at High Flow Velocities. M.S., 7/79. Advisor: S. C. Jain.

French, Mark N. Quantitative Real-Time Rainfall Forecasting Using Remote Sensing. Ph.D., 8/92. Advisor: W. F. Krajewski.

Frisbie, Todd Russell. Measurements of Plow Loads During Ice Removal Operations from Pavement. M.S., 12/94. Advisor: W. A. Nixon.

Fukuoka, Shoji. Longitudinal Dispersion in Sinuous Channels. Ph.D., 1/71. Advisor: W. W. Sayre.

Fuller, Robert Brady. Nitrogen Modeling in One-Dimensional Unsteady Flow. M.S., 5/96. Advisor: F. M. Holly, Jr.

Gawronski, Tomasz Jakub. Blade Geometry Effects on Ice Scraping Forces. M.S., 8/94. Advisor: W. A. Nixon.

Gay, George E. Model Study of the Quad-Cities Nuclear Power Plant Intake Structure. M.S., 5/79. Advisors: T. Nakato and W. W. Sayre.

Giaquinta, Arthur Ralph. Numerical Modelling of Unsteady Flow with Natural and Forced Separation. Ph.D., 5/74. Advisor: E. O. Macagno.

Gogus, Mustafa. Flow Characteristics below Floating Covers with Application to Ice Jams. Ph.D., 12/80. Advisor: J. C. Tatinclaux.

Goodman, Brett P. Comprehensive Dissolved Oxygen Modeling in One-Dimensional Unsteady Flow. M.S., 12/96. Advisor: F. M. Holly, Jr.

Grecu, Mircea. Quality Control of Weather Radar Echo Using Neural Networks. M.S., 7/96. Advisor: W. F. Krajewski.

Grimm-Strele, Jost. Longitudinal Mixing of Heated Water in Open-Channel Flow. Ph.D., 12/75. Advisor: W. W. Sayre.

Guetter, Alexandre K. An Analytical Solution for Stratified Flows Induced by Settling Solids. M.S., 8/88. Advisor: S. C. Jain.

Guetter, Alexandre K. Hydrology of the Continental United States. Ph.D., 12/93. Advisor: K. P. Georgakakos.

Guo, Zhi. General Planar Translation of Two Bodies in a Fluid. Ph.D., 5/90. Advisor: A. T. Chwang.

Güven, Oktay. An Experimental and Analytical Study of Surface-Roughness Effects on the Mean Flow past Circular Cylinders. Ph.D., 12/75. Advisors: V. C. Patel and C. Farell.

Haferman, Jeffrey Lawrence. A Polarized Multi-Dimensional Discrete-Ordinates Radiative Transfer Model for Remote Sensing Applications. Ph.D., 12/95. Advisors: T. F. Smith and W. F. Krajewski.

Hamdy, Usama Mohamed Aly. A Damage-Based Life Prediction Model of Concrete under Variable Amplitude Fatigue Loading. Ph.D., 5/97. Advisor: W. A. Nixon.

Han, Taeyoung. A Flow-Visualization Study of Three-Dimensional Boundary-Layer Separation on Bodies of Revolution at Incidence. M.S., 5/77. Advisor: V. C. Patel.

Haniu, Hiroyuki. An Experimental Study of Two-Dimensional Buoyant Jets in Cross-Flow. M.S., 7/79. Advisor: B. R. Ramaprian.

Haniu, Hiroyuki. Turbulence Measurements in Plane Jets and Plumes in Cross-Flow. Ph.D., 7/83. Advisor: B. R. Ramaprian.

Hatfield, Kirk. Development of Steady-State Regional Stream Model for the Aquatic Partitioning and Transport of Toxic Substances. M.S., 5/82. Advisor: J. L. Schnoor.

Hayden, Warren Scott. Hydraulic Characteristics of a Truncated Helicoidal-Ramp Dropstructure. M.S., 8/93. Advisor: S. C. Jain.

Henne, Randall Lee. Scale Effects on Oxygen Transfer in Bubbly Turbulent Shear Flow. M.S., 5/93. Advisor: S. C. Jain.

Hewlett, Ben Yao. Rate of Recession of the Leading Edge of Ice Covers on Open Channel Flows. M.S., 7/76. Advisor: G. Dagan and E. O. Macagno.

Hirayama, Ken-Ichi. An Investigation of Ice Forces on Vertical Structures. Ph.D., 5/74. Advisor: J. Schwarz.

Ho, Kuo-San. Finite Analytic Numerical Method for Laminar Two-Dimensional Flow and Heat Transfer Problems Using Boundary Fitted Coordinates. Ph.D., 5/83. Advisor: C.-J. Chen.

Hoeksema, Robert James. The Geostatistical Approach to the Inverse Problem in Two-Dimensional Steady State Groundwater Modeling. Ph.D., 12/84. Advisor: P. K. Kitanidis.

Hsu, Chung-Chieh. A Sediment-Budget Analysis for the Upper Mississippi River between Guttenberg, Iowa, and Saverton, Missouri. M.S., 12/82. Advisors: T. Nakato and J. F. Kennedy.

Hsu, Chung-Chieh. Mechanics of Flow in Alluvial River Bends with Application to Iowa-Vane Bank Protection. Ph.D., 5/87. Advisor: F. M. Holly, Jr.

Hsu, Kuang-Shen. Diffusion of Polymers in a Developing Boundary Layer. M.S., 1/71. Advisor: L. Landweber.

Hsu, Kuang-Shen. Spectral Evolution of Ice Ripples. Ph.D., 12/73. Advisors: J. F. Kennedy and F. Locher.

Hsu, Min-Chiang. Deformation and Stress Analysis of the Left Ventricle: An Active Thick Wall Model. M.S., 5/80. Advisor: C.-J. Chen.

Hsu, Pei-Pei. Added Moment of Inertia of Two-Dimensional Sections at a Free Surface. M.S., 8/80. Advisor: L. Landweber.

Hsu, Pei-Pei. Centerplane Source Distribution for Slender Bodies. Ph.D., 5/87. Advisor: L. Landweber.

Hsu, Shaohua Marko. An Assessment of Analytical and Numerical Prediction of One-Dimensional Mobile-Bed Dynamics. Ph.D., 12/91. Advisor: F. M. Holly, Jr.

Hsu, Sheng-Chuan. Mathematical Modelling of Flow and Bed Characteristics in Curved Channels with Nonuniform Bed Material. Ph.D., 8/88. Advisor: A. J. Odgaard.

Hsu, Tai-Dan. Optimal Design of Wet Tower/Once-Through Hybrid Cooling Systems. Ph.D., 12/78. Advisors: T. E. Croley II and A. R. Giaquinta.

Hsueh, Nai-Drwang. An Evaluation of Movable Bed Tidal Inlet Models. M.S., 7/77. Advisor: S. C. Jain.

Huang, Ching-Jer. Diffraction of Acoustic Waves by a Ring Aperture in a Baffle of Arbitrary Impedance. Ph.D., 5/91. Advisor: A. T. Chwang.

Huang, Hanping. Effects of Waves and Free Surface on Turbulence in the Boundary Layer of a Surface-Piercing Flat Plate. M.S., 8/94. Advisor: F. Stern.

Huang, Hung-Pin. Ice Formation in Frequently Transited Navigation Channels. Ph.D., 8/88. Advisor: R. Ettema.

Huang, Liang-Hsiung. Rotating Oblate Bodies in a Viscous Fluid. M.S., 12/83. Advisor: A. T. Chwang.

Huang, Liang-Hsiung. Trapping and Absorption of Underwater Sound. Ph.D., 12/86. Advisor: A. T. Chwang.

Huang, Long-Cheng. An Analytical and Experimental Study of Multiple Plumes. M.S., 12/79. Advisor: S. C. Jain.

Huang, Long-Cheng. Seismic Water Pressures on Dams for Arbitrarily Shaped Reservoirs. Ph.D., 8/84. Advisor: A. T. Chwang.

Huang, Tseng-Hsiang. Earthquake Effects on Rectangular Dam-Reservoir Systems. M.S., 12/80. Advisor: A. T. Chwang.

Hwang, Guang-Jiunn. Buoyancy Effects in Thermally Stratified Open-Channel Flow. Ph.D., 7/75. Advisor: W. W. Sayre.

Hwang, Jieh-Chyuan. Investigation and Evaluation of the Thermal Discharge Systems of Dresden Nuclear Station. M.S., 5/76. Advisor: W. W. Sayre.

Hwang, Robert Rong-Jiann. On the Hydrodynamic Stability of Stoke's First and Second Problems. Ph.D., 5/74. Advisor: C.-J. Chen.

Hwang, Wei-Shien. Rotating Flows of a Viscous Fluid. Ph.D., 8/89. Advisor: A. T. Chwang.

Hyun, Beom-Soo. Measurements in the Flow around a Marine Propeller at the Stern of an Axisymmetric Body. Ph.D., 5/90. Advisor: V. C. Patel.

Jain, Subhash Chandra. Evolution of Sand Wave Spectra. Ph.D., 1/71. Advisor: J. F. Kennedy.

Jaramillo-Torres, Wilson F. Aggradation and Degradation of Alluvial-Channel Beds. Ph.D., 5/83. Advisor: S. C. Jain.

Johnson, Jon Kent. The Iowa Impulse Fish Guidance System: A Microcomputer-Controlled Fish Guidance System Feasibility Study. Ph.D., 12/86. Advisor: D. B. McDonald.

Johnson, Timothy A. Visualization of Topology of Separated Flow over a Semi-Elliptic Wing at Incidence Intersecting a Plane Wall. M.S., 12/91. Advisor: V. C. Patel.

Johnson, Timothy A. Numerical and Experimental Investigation of Flow past a Sphere up to a Reynolds Number of 300. Ph.D., 12/96. Advisor: V. C. Patel.

Jovic, Srboljub D. Large-Scale Structure of the Turbulent Wake behind a Flat Plate. Ph.D., 5/86. Advisor: B. R. Ramaprian.

Ju, Sangseon. Study of Total and Viscous Resistance for the Wigley Parabolic Ship Form. M.S., 5/83. Advisor: L. Landweber.

Ju, Sangseon. Numerical Study of Ship Stern and Wake Flows at Model and Full-Scale Reynolds Numbers. Ph.D., 8/89. Advisor: V. C. Patel.

Jun, Kyung Soo. Oxygen Transfer in Closed-Conduit Turbulent Shear Flows. Ph.D., 8/91. Advisor: S. C. Jain.

Kadle, Durgaprasad Shripad. Hydrodynamic Effect of Earthquakes on Circular Dam-Reservoir Systems. M.S., 7/81. Advisor: A. T. Chwang.

Kalale, Krishnaprasad L. Development of a Heat-Tagging Technique for the Study of Large-Scale Mixing in a Developing Two-Dimensional Wake. M.S., 5/84. Advisor: B. R. Ramaprian.

Kang, Shin-Hyoung. Viscous Effects on the Wave Resistance of a Thin Ship. Ph.D., 7/78. Advisor: L. Landweber.

Kao, Erh-Ying. No thesis. M.S., 12/91.

Karathanasi, Irene. No thesis. M.S., 7/83. Advisor: S. C. Jain.

Karim, Md. Fazle. Computer-Based Predictors for Sediment Discharge and Friction Factor of Alluvial Streams. Ph.D., 12/81. Advisor: J. F. Kennedy.

Keng, Titus T. C. Thermal Regimes of the Mississippi and Missouri Rivers Downstream from the Southern Iowa Border. M.S., 5/78. Advisors: A. R. Giaquinta and R. R. Dague.

Kessey, Marion E. Armoring and Pavement in Gravel-Bed Rivers: Numerical and Physical Analysis. Ph.D., 5/93. Advisor: S. C. Jain.

Khalighi, Bahram. Numerical Solution of Two Dimensional Poisson and Laplace Equations by Finite Analytic Methods. M.S., 5/80. Advisor: C.-J. Chen.

Kim, Hyoung-Tae. Computation of Viscous Flow around a Propeller-Shaft Configuration with Infinite-Pitch Rectangular Blades. Ph.D., 5/89. Advisor: F. Stern.

Kim, Sungeun. Numerical Studies of Three-Dimensional Flow Separation. Ph.D., 5/91. Advisor: V. C. Patel.

Kim, Wu Joan. An Experimental and Computational Study of Longitudinal Vortices in Turbulent Boundary Layers. Ph.D., 12/91. Advisor: V. C. Patel.

Kouame, Nguessan. No thesis. M.S., 12/83. Advisor: A. J. Odgaard.

Krishan, Dharmvir. Shallow Flows over Large Roughness Elements. M.S., 12/84. Advisor: S. C. Jain.

Kulkarni, Manohar. No thesis. M.S., 12/80. Advisor: B. R. Ramaprian.

Kumar, Sree. Incorporation of the SCS Infiltration Model into the IIHR Watershed Model. M.S., 8/80. Advisor: S. C. Jain.

Lakshmi, Venkataraman. Investigation of the Effects of Rainfall Input Variability Using a Distributed Parameter Catchment Model. M.S., 8/89. Advisor: W. F. Krajewski.

Lan, Paul Bo. Analysis of Degradation and Aggradation of Missouri River between Omaha and Nebraska City. M.S., 12/87. Advisors: T. Nakato and S. C. Jain.

Lazaro, Javier. Dynamics of Continuous Mode Icebreaking. M.S., 5/87. Advisors: R. Ettema and F. Stern.

Lee, Chu-Liang. Ice Jam Initiation by Partial Surface Obstructions. M.S., 5/76. Advisor: J. C. Tatinclaux.

Lee, Chu-Liang. Flow and Absorption in a Contracting Channel with Application to the Human Intestine. Ph.D., 7/81. Advisor: E. O. Macagno.

Lee, Dae Un. No thesis. M.S., 7/76. Advisor: J. G. Melville.

Lee, Hong-Yuan E. Flow and Bed Characteristics in Alluvial Channel Bends. Ph.D., 5/84. Advisor: A. J. Odgaard.

Lee, Hsai-Yin. Prediction of Transient Surface Heat Flux and Temperature on a Hollow Cylinder. M.S., 5/81. Advisor: C.-J. Chen.

Lee, Jae-Soo. Modeling and Analysis of Soil Moisture Dynamics at Climate Scales. Ph.D., 5/93. Advisor: W. F. Krajewski.

Lee, Rosa Ming-Hsing. Optimum Combination of Selected Cooling Alternatives for Electric Power Plants. Ph.D., 12/78. Advisors: T. E. Croley II and A. R. Giaquinta.

Lee, Tim Hau. Simulation Studies of the Inverse Problem in One-Dimensional Unsteady Groundwater Flow. M.S., 5/86. Advisor: K. P. Georgakakos.

Lee, Tim Hau. A Stochastic-Dynamical Model for Short-Term Quantitative Rainfall Prediction. Ph.D., 5/91. Advisor: K. P. Georgakakos.

Lee, Yih-Fey. Meandering of Supraglacial Channels. M.S., 7/79. Advisor: J. C. Tatinclaux.

Lee, Yu-Tai. Two Blockage-Effect Studies: Wind Tunnel Measurements on a Prolate Spheroid and a Mathematical Model of the Flow about a Normal Flat Plate with Separation. M.S., 12/75. Advisor: L. Landweber.

Lee, Yu-Tai. Thick Axisymmetric Turbulent Boundary Layer and Wake of a Low-Drag Body. Ph.D., 12/78. Advisor: V. C. Patel.

Lehman, Roger William. Calibration and Adaptation of Iowa Rapid Sediment Analyzer to an IBM Compatible Microcomputer. M.S., 5/88. Advisors: T. Nakato and S. C. Jain.

Li, Peter. Finite Differential Methods—Applications of Analytical Solution Techniques to the Numerical Solutions of Partial Differential Equations. Ph.D., 7/78. Advisor: C.-J. Chen.

Life, Kymberly Jo. No thesis. M.S., 8/84. Advisor: C.-J. Chen.

Lin, Dah-Syang. An Application of Recursive Methods to Radar-Rainfall Estimation. M.S., 12/89. Advisor: W. F. Krajewski.

Lin, Fangbiao. Development of a Numerical Method for Three-Dimensional Incompressible Flow with Multigrid Acceleration and Near-Wall Turbulence Closure. Ph.D., 7/96. Advisors: A. J. Odgaard, V. C. Patel, and F. Sotiropoulos.

Lin, Guoliang. A New Method for Evaluation of Monthly Rainfall over Open Ocean Using Underwater Acoustic Sensors. M.S., 8/93. Advisor: W. F. Krajewski.

Lin, Hsiao-Wen. A Control-Volume Finite-Element Method for Navier-Stokes Equations. M.S., 8/89. Advisors: V. C. Patel and L.-D. Chen.

Lin, Shyi-Jang. Instability of Two Dimensional Incompressible Parallel Turbulent Shear Flow. M.S., 7/78. Advisor: C.-J. Chen.

Liong, Shie-Yui. Effects of Vertical Distortion on Thermal-Hydraulic Modeling of Surface Discharges. Ph.D., 7/77. Advisor: W. W. Sayre.

Liu, Connie Chunyan. Numerical Analysis of Error Quantification and Propagation in Radar Remote Sensing for Rainfall Estimation. Ph.D., 12/95. Advisor: W. F. Krajewski.

Lo, Chien-Kuo. Risk Consideration in Reservoir Operation Optimization with Application to the Red Rock Reservoir, Iowa. Ph.D., 12/81. Advisor: P. K. Kitanidis.

Longo, Joseph Frank. Scale Effects on Near-Field Wave Patterns. M.S., 12/90. Advisor: F. Stern.

Longo, Joseph Frank. Effects of Yaw on Model-Scale Ship Flows. Ph.D., 5/96. Advisor: F. Stern.

Lucie, Christopher David. Laboratory Experiments on Ice Jams in River Bends. M.S., 5/89. Advisor: R. Ettema.

Luknanto, Djoko. Numerical Simulation of Saturated Groundwater Flow and Pollutant Transport in Karst Regions. Ph.D., 12/91. Advisor: F. M. Holly, Jr.

Luzbetak, Douglas Joseph. Piping as a Mechanism of Bank Erosion along Rivers in Iowa. M.S., 12/87. Advisor: A. J. Odgaard.

Maisch, Federico. Wind Loading on Hyperbolic Cooling Towers. M.S., 7/74. Advisor: C. Farell.

Mannheim, Carl Olaf Magnus. A Unique Approach of Modeling Gas Supersaturation Using a Physical Model. M.S., 5/97. Advisor: L. Weber.

McCollum, Jeffrey Richard. Studies of Calibration Procedures for Space-Based Rainfall Estimation Methods. M.S., 8/94. Advisor: W. F. Krajewski.

McCollum, Jeffrey Richard. Error Analysis of Global Rainfall Estimation Algorithms. Ph.D., 8/97. Advisor: W. F. Krajewski.

McEnroe, Bruce M. Two-Dimensional Heated Jet in Shallow Stagnant Water. M.S., 7/78. Advisor: S. C. Jain.

McKechnie, Deborah Jean. Transport of Soluble Pollutants in a Run-of-the-River Impoundment. Ph.D., 8/88. Advisors: J. L. Schnoor and F. M. Holly, Jr.

Menendez, Angel Nicolas. Study of Unsteady Turbulent Boundary Layers. Ph.D., 12/83. Advisor: B. R. Ramaprian.

Merino, Marcelo P. Internal Shear Strength of Floating Fragmented Ice Covers. M.S., 5/74. Advisor: J. F. Kennedy.

Meselhe, Ehab Amin. Simulation of Unsteady Flow in Irrigation Canals with Dry Bed. M.S., 12/91. Advisor: R. Ettema.

Meselhe, Ehab Amin. Numerical Simulation of Transcritical Flow in Open Channels. Ph.D., 5/94. Advisor: F. M. Holly, Jr.

Miloh, Touvia. Higher-Order Theory of Ship Waves from Centerplane Source Distributions. Ph.D., 1/72. Advisor: L. Landweber.

Moran, David Dunstan. Finite-Integral Longitudinal-Cut Technique for Computing the Wave Resistance of a Ship Model in a Towing Tank. Ph.D., 5/71. Advisor: L. Landweber.

Moreno, Miguel. Experimental Study of the Influence of the Wake on the Wavemaking Resistance of a Ship Model. M.S., 12/75. Advisor: L. Landweber.

Mosconi, Carlos E. Effect of Streamline Curvature on Friction Factor. M.S., 12/84. Advisor: S. C. Jain.

Mosconi, Carlos E. River-Bed Variations and Evolution of Armor Layers. Ph.D., 5/88. Advisor: S. C. Jain.

Mullusky, Mary G. Sensitivity of Large-Basin Hydrology, Forecasts and Management to Historical Climatic Forcing. M.S., 8/93. Advisor: K. P. Georgakakos.

Munukutla, Sastry S. Turbulent Wake Development behind Streamlined Bodies. Ph.D., 5/81. Advisors: B. R. Ramaprian and V. C. Patel.

Musgrove, Daniel D. A Two-Dimensional Suspended Solids Model for Dredge Disposal on the Mississippi River. M.S., 5/80. Advisor: J. L. Schnoor.

Muste, Marian Valer-Ioan. Particle and Liquid Velocity Measurements in Sediment-Laden Flows with a Discriminator Laser-Doppler Velocimeter. Ph.D., 8/95. Advisors: V. C. Patel and R. N. Parthasarathy.

Nakato, Tatsuaki. Wave-Induced Sediment Entrainment from Rippled Beds. Ph.D., 12/74. Advisors: F. A. Locher and J. F. Kennedy.

Nakayama, Akihiko. Viscid-Inviscid Interaction Due to the Thick Boundary Layer near the Tail of a Body of Revolution. Ph.D., 12/74. Advisors: V. C. Patel and L. Landweber.

Naseri-Neshat, Hamid. Development of the Finite-Differential Numerical Method and Its Application to Two-Dimensional Navier-Stokes Equation. Ph.D., 12/79. Advisor: C.-J. Chen.

Neary, Vincent Sinclair. Flow Structure at an Open Channel Diversion. M.S., 8/92. Advisor: A. J. Odgaard.

Neary, Vincent Sinclair. Numerical Modeling of Diversion Flows. Ph.D., 12/95. Advisors: A. J. Odgaard and F. Sotiropoulos.

Nikitopolous, Constantinos P. Buoyancy Effects in the Zone of Flow Establishment of Round Vertical Buoyant Jets Discharged into a Stagnant Environment. M.S., 5/78. Advisor: S. C. Jain.

Noblesse, Francis. Unsteady Nonuniform Flow in the Entrance of a Pipe. M.S., 5/71. Advisor: C. Farell.

Noblesse, Francis. A Perturbation Analysis of the Wavemaking of a Ship, with an Interpretation of Guilloton's Method. Ph.D., 12/74. Advisor: L. Landweber.

Nospal, Aleksandar. Investigation of the Thermal Near-Field for an Alternating Multiport Diffuser Pipe. M.S., 12/74. Advisor: J. C. Tatinclaux.

Novak, Charles J. Measurements in the Wake of an Infinite Swept Airfoil. M.S., 12/81. Advisor: B. R. Ramaprian.

Obasih, Kemakolam. Finite Analytic Numerical Solution of Heat Transfer for Flow past a Rectangular Cavity. M.S., 6/81. Advisor: C.-J. Chen.

Obasih, Kemakolam. Prediction of Laminar and Turbulent Flows past Single and Twin Airfoils. Ph.D., 7/87. Advisor: C.-J. Chen.

Onishi, Yasuo. Effects of Meandering on Sediment Discharges and Friction Factors of Alluvial Streams. Ph.D., 12/72. Advisor: J. F. Kennedy.

Paez, Diana. Experimental Study of Truncated Helicoidal-Ramp Drop Structures. M.S., 5/92. Advisor: S. C. Jain.

Paily, Poothrikka P. Winter-Regime Thermal Response of Heated Streams. Ph.D., 5/74. Advisor: E. O. Macagno.

Paredes-Tasan, Edison. No thesis. M.S., 12/75. Advisor: T. E. Croley II.

Park, Inbo. Numerical Simulation of Aggradation and Degradation of Alluvial-Channel Beds. Ph.D., 5/87. Advisor: S. C. Jain.

Parr, Alfred David. Prototype and Model Studies of the Diffuser-Pipe System for Discharging Condenser Cooling Water at the Quad Cities Nuclear Power Station. Ph.D., 5/76. Advisor: W. W. Sayre.

Paterson, Eric George. Computation of Natural and Forced Unsteady Viscous Flow with Application to Marine Propulsors. Ph.D., 5/94. Advisor: F. Stern.

Peña, Jose Maria. On the Behavior of Warm, Sinking Jets Discharged into Water of Low Temperature. M.S., 5/74. Advisor: J. C. Tatinclaux.

Perez-Rojas, Luis. Effect of Surface Roughness on the Viscous Resistance of a Ship Model Determined by a Wake Survey. M.S., 12/75. Advisor: L. Landweber.

Pokorny, Michael Jon. Video Based Data-Acquisition Technique for Measuring Oscillatory Loads on a Galloping Conductor. M.S., 8/93. Advisor: R. Ettema.

Potter, James Daniel. Measurement of Ice Scraping Forces on Snow-plow Underbody Blade. M.S., 5/96. Advisor: W. A. Nixon.

Power Meneses, Henry. On Cylindrical Solitary Waves. Ph.D., 5/81. Advisor: A. T. Chwang.

Powers, Andrew. No thesis. M.S., 1/71.

Pujol, Alfonso. Numerical Experiments on the Stability of Poiseuille Flows of Non-Newtonian Fluids. Ph.D., 8/71. Advisor: E. O. Macagno.

Rajagopal, Srinivasan. Internal Hydraulic Jumps with Mixing. M.S., 5/71. Advisor: E. O. Macagno.

Rajaram, Harihar. Recursive Parameter Estimation of Conceptual Watershed Response Models. M.S., 12/87. Advisor: K. P. Georgakakos.

Raja Rao, Kuchibhotla. Stochastic Trade-Offs for Reservoir Operation. Ph.D., 12/76. Advisor: T. E. Croley II.

Richmond, Marshall Charles. Surface Curvature and Pressure Gradient Effects on Turbulent Flow: An Assessment Based on Numerical Solution of the Reynolds Equations. Ph.D., 7/87. Advisor: V. C. Patel.

Robison, Charles Price. A Computer Model for Predicting Suspended-Sediment Plumes from GREAT III Dredge-Disposal Operations. M.S., 7/82. Advisors: J. Schnoor and A. R. Giaquinta.

Román Gallardo, Marco A. Influence of Air and Water Conditions on the Melting Rate of Ice Covers. M.S., 7/74. Advisors: E. O. Macagno and J. Schwarz.

Sanchez Aliás, Alejo. A General Purpose Radiative Transfer Model for Application of Remote Sensing in Multi-Dimensional Systems. Ph.D., 12/91. Advisor: W. F. Krajewski.

Sancholuz, Arturo G. A Statistical Study of the Spike Bursts on the Slow Wave in the Duodenum. M.S., 7/74. Advisors: J. R. Glover and E. O. Macagno.

Santos, Rafael. No thesis. M.S., 8/90. Advisor: C.-J. Chen.

Sarda, Om Prakash. No thesis. M.S., 12/80. Advisor: J. F. Kennedy.

Sarda, Om Prakash. Turbulent Flow past Ship Hulls—An Experimental and Computational Study. Ph.D., 8/86. Advisor: V. C. Patel.

Satija, Kanwar Sain. On the Thick Boundary Layer near the Tail of a Body of Revolution. Ph.D., 1/71. Advisor: L. Landweber.

Sato, Chikashi. Characterization of Heavy Metals Effects on Nitrosomonas Europea by Graphical and Regression Methods. Ph.D., 5/81. Advisor: D. B. McDonald.

Sauvaget, Patrick. A Numerical Model for Stratified Flows in Estuaries and Reservoirs. M.S., 8/85. Advisor: F. M. Holly, Jr.

Savic, Ljubodrag. Computation of Open-Channel Discontinuous Flows Using the Modified Godunov Method. Ph.D., 5/91. Advisor: F. M. Holly, Jr.

Schiller, Eric John. Vertical Mixing of Heated Effluents in Open-Channel Flow. Ph.D., 7/73. Advisor: W. W. Sayre.

Schmidt, Charles Chris. Development and Evaluation of a Laser Doppler Velocimeter System for Measuring Frazil Ice Concentration. M.S., 7/74. Advisor: J. R. Glover.

Schohl, Gerald Allen. Naled Ice Growth. Ph.D., 12/85. Advisor: R. Ettema.

Seki, Masakazu. Inter-District Water Allocations via Linear Programming and Linear Programming Decomposition. M.S., 7/77. Advisor: T. E. Croley II.

Sencial, Circe Urania. No thesis. M.S., 5/75. Advisor: W. W. Sayre.

Serre, Mark. A Study of Energy Loss at Combining Pipe Junction in Fish Bypass Systems. M.S., 5/92. Advisor: A. J. Odgaard.

Shahshahan, Ali. Study of Free-Surface Flow near a Ship Bow. M.S., 7/81. Advisor: L. Landweber.

Shahshahan, Ali. Effects of Viscosity on Wavemaking Resistance of a Ship Model. Ph.D., 8/85. Advisor: L. Landweber.

Sharifi, Mohammad Bagher. Chaotic Dynamic Analysis of Natural Phenomena: Convective Storm Rainfall. Ph.D., 12/90. Advisor: K. P. Georgakakos.

Sheikholeslami, Mohamad Zahed. Application of Finite Analytic Method to the Numerical Solution of Two-Point Boundary Value Problems of Ordinary Differential Equations. M.S., 5/80. Advisor: C.-J. Chen.

Sheikholeslami, Mohamad Zahed. Computer Aided Analysis of Two-Dimensional Laminar and Turbulent Incompressible Flows. Ph.D., 5/86. Advisor: C.-J. Chen.

Shen, Huiying. Spatially-Lumped Precipitation Prediction Models for Real-Time Hydrology. M.S., 5/89. Advisor: K. P. Georgakakos.

Shen, Hung-Tao. Second-Order Wave Resistance Theory of a Submerged Spheroid. Ph.D., 7/74. Advisor: C. Farell.

Shie, Kuo-Fu. Three-Dimensional Modeling of Buoyant Surface Discharge. M.S., 12/81. Advisor: A. J. Odgaard.

Shih, Kuo-Kung. The Impact of Urbanization on Floods of Ralston Creek. Ph.D., 12/78. Advisor: T. E. Croley II.

Silva, Jose Matos. A Numerical Model of Bed Degradation along the Missouri River between Yankton (SD) and Omaha (NE). M.S., 7/82. Advisor: J. F. Kennedy.

Silva, Jose Matos. Kinematic-Wave Analysis of River Responses to Imposed Disequilibria. Ph.D., 12/86. Advisor: J. F. Kennedy.

Singerman, Robert B. J. Fluid Mechanics of the Human Duodenum. Ph.D., 5/74. Advisor: E. O. Macagno.

Singh, Kanwerdip. Finite Analytic Numerical Solutions of Two-Dimensional Navier-Stokes Equations in Primitive Variables. M.S., 7/81. Advisor: C.-J. Chen.

Singh, Kanwerdip. Development of a Two-Scale Turbulence Model and Its Applications. Ph.D., 5/85. Advisor: C.-J. Chen.

Sinha, Sanjiv Kumar. Three-Dimensional Numerical Model for Turbulent Flows through River Reaches of Complex Bathymetry. Ph.D., 5/96. Advisors: A. J. Odgaard and F. Sotiropoulos.

Sirviente, Ana Isabel. No thesis. M.S., 12/91. Advisor: V. C. Patel.

Sirviente, Ana Isabel. Wake of an Axisymmetric Body Propelled by a Jet with and without Swirl. Ph.D., 7/96. Advisor: V. C. Patel.

Sium, Ogbazghi. Transverse Flow Distribution in Natural Streams as Influenced by Cross-Sectional Shape. M.S., 7/75. Advisor: W. W. Sayre.

Smith, Brennan Thomas. No thesis. M.S., 12/92. Advisor: F. M. Holly, Jr.

Smith, Brennan Thomas. Ice-Cover Influence on Flow and Bedload Transport in Dune-Bed Channels. Ph.D., 5/95. Advisor: R. Ettema.

Song, Gang Bog. Sediment Transport and Bed Forms under Ice Covers. M.S., 5/78. Advisor: W. W. Sayre.

Spasojevic, Miodrag. Numerical Simulation of Two-Dimensional (Plan View) Unsteady Water and Sediment Movement in Natural Watercourses. Ph.D., 12/88. Advisor: F. M. Holly, Jr.

Spoljaric, Anita. Mechanics of Submerged Vanes on Flat Boundaries. Ph.D., 5/88. Advisor: A. J. Odgaard.

Srinivasan, P. Numerical Solution of Two-Dimensional Transverse Diffusion Equation. M.S., 12/81. Advisors: A. R. Giaquinta and J. L. Schnoor.

Stavitsky, David. Flow and Mixing in a Contracting Channel with Applications to the Human Intestine. Ph.D., 5/79. Advisor: E. O. Macagno.

Steurer, Tracy Lynn. A GIS-Based Study of Sediment Transport in the North Branch of Ralston Creek Using the ANSWERS Models. M.S., 5/96. Advisor: F. Weirich.

Sturm, Terry W. An Analytical and Experimental Investigation of Density Currents in Sidearms of Cooling Ponds. Ph.D., 5/76. Advisor: J. F. Kennedy.

Su, Tzuoh-Ying. Thermal Regimes of Upper Mississippi and Missouri Rivers and Hybrid Once-Through Wet Tower Cooling Systems for Power Plants. Ph.D., 5/77. Advisor: J. F. Kennedy.

Subramani, Anil Kumar. Extensions of a Viscous Ship-Flow CFD Method for the Calculation of Sinkage and Trim. M.S., 12/96. Advisor: F. Stern.

Sudjarwadi. A Simulation Model of Irrigation Water Requirements for Paddies in the Mountainous Progo River Basin. Ph.D., 12/86. Advisor: K. P. Georgakakos.

Tahara, Yusuke. An Interactive Approach for Calculating Ship Boundary Layers and Wakes for Nonzero Froude Number. Ph.D., 8/92. Advisor: F. Stern.

Tan, Cheng-Ann. Numerical Simulation of Contaminant Mixing in Ice-Covered Channels. M.S., 12/97. Advisors: R. Ettema and S. Sinha.

Tang, Chii-Jau. The Sensitivity of Boussinesq Approximation to Buoyant Jet Characteristics. M.S., 12/81. Advisor: S. C. Jain.

Tang, Chii-Jau. Free-Surface Flow Phenomena Ahead of a Two-Dimensional Body in a Viscous Fluid. Ph.D., 5/87. Advisors: L. Landweber and V. C. Patel.

Tao, Men-Cheh. Turbulent Mixing of Density Stratified Fluids. Ph.D., 1/71. Advisor: C. Farell.

Tapia-Aviles, Felix. No thesis. M.S., 12/78. Advisor: T. E. Croley II.

Teal, Martin Joseph. Estimation of Mean Velocity for Flow under Ice Cover. M.S., 5/93. Advisor: R. Ettema.

Tembhekar, Shekhar. No thesis. M.S., 12/81. Advisor: S. C. Jain.

Toda, Keiichi. Numerical Modelling of Advection Phenomena. Ph.D., 5/86. Advisor: F. M. Holly, Jr.

Trejos, Saul. No thesis. M.S., 5/88. Advisor: S. C. Jain.

Tsai, Chii-ell. Study of Total, Viscous and Wave Resistance of a Family of Series-60 Models; Further Development of a Procedure for Determination of Wave Resistance from Longitudinal-Cut, Surface-Profile Measurements. Ph.D., 12/72. Advisor: L. Landweber.

Tsai, Ping-Ho. Turbulent Flow in a Curved Streamwise Corner. M.S., 8/85. Advisor: V. C. Patel.

Tsai, Whey-Fone. A Study of Ice-Covered Bend Flow. Ph.D., 8/91. Advisor: R. Ettema.

Tsou, I. No thesis. M.S., 12/86. Advisor: A. J. Odgaard.

Tu, Shuen-Wei. An Experimental Study of Oscillatory Pipe Flow at Transitional Reynolds Number. M.S., 7/78. Advisor: B. R. Ramaprian.

Tu, Shuen-Wei. Study of Periodic Turbulent Pipe Flow. Ph.D., 12/81. Advisor: B. R. Ramaprian.

Tyan, Goan-Liau. Evaluation of Some Sediment-Transport Formulas for a Reach of Pool 20 of the Mississippi River. M.S., 12/78. Advisor: T. Nakato and W. W. Sayre.

Tyan, Wen Hua. No thesis. M.S., 5/86. Advisor: C.-J. Chen.

Tyndall, Juliet. A Numerical Study of Flow over Wavy Walls. M.S., 12/88. Advisor: V. C. Patel.

Tziara, Magdaleni. No thesis. M.S., 7/83. Advisor: S. C. Jain.

Urroz-Aguirre, Gilberto. Studies on Ice Jams in River Bends. Ph.D., 5/88. Advisor: R. Ettema.

Uzuner, Ahmet Selcuk. Optimum Mechanical and Natural Draft Wet Cooling Towers. M.S., 5/75. Advisors: V. C. Patel and T. E. Croley II.

Uzuner, Mehmet Secil. Stability of Floating Ice Blocks. M.S., 5/71. Advisor: J. F. Kennedy.

Uzuner, Mehmet Secil. Hydraulics and Mechanics of River Ice Jams. Ph.D., 5/74. Advisor: J. F. Kennedy.

Vadnal, John Louis. Field Study of Sediment Transport Characteristics in the Mississippi River near Buzzard Island. M.S., 12/79. Advisors: T. Nakato and J. F. Kennedy.

Vadnal, John Louis. A Numerical Model for Steady Flow in Meandering Alluvial Channels. Ph.D., 12/84. Advisors: T. Nakato and J. F. Kennedy.

Viswesvaran, Chockalingam. Sensitivity Analysis of an Integrated Hydro-meteorological Model. M.S., 12/88. Advisor: K. P. Georgakakos.

Vomvoris, Efstratios G. Groundwater Parameter Estimation: A Geostatistical Approach. M.S., 7/82. Advisor: P. K. Kitanidis.

Walker, William Karl. Continuous Long-Term Simulation to Determine the Frequency and Severity of Extreme River Temperatures. M.S., 5/96. Advisor: A. A. Bradley.

Walter, Joel Allan. Development of Quantitative Velocity Imaging System and Its Application to Offset Channel Flow. M.S., 12/89. Advisor: C.-J. Chen.

Walter, Joel Allan. Three-Dimensional Wake of a Surface-Mounted Ellipsoid. Ph.D., 5/97. Advisor: V. C. Patel.

Wang, Hsio-Min. No thesis. M.S., 5/86. Advisor: C.-J. Chen.

Wang, Keh-Han. The Shear Strength of Mush Ice—An Experimental Study. M.S., 5/81. Advisor: J. C. Tatinclaux.

Wang, Keh-Han. Nonlinear Impulsive Motion of a Vertical Cylinder. Ph.D., 12/85. Advisor: A. T. Chwang.

Wang, Tzuu Po. Thickness of Ice Jam Due to Accumulation and Submergence of Ice Floes. M.S., 5/76. Advisor: J. C. Tatinclaux.

Wang, Yalin. Sediment Control with Submerged Vanes. Ph.D., 5/91. Advisor: A. J. Odgaard.

Wasif, Mir Alimuddin. Compressive Behavior of S2 Freshwater Ice. M.S., 8/92. Advisor: W. A. Nixon.

Wasimi, Saleh Ahmed. Operation of a System of Reservoirs under Flood Conditions Using Linear Quadratic Gaussian Stochastic Control. Ph.D., 7/83. Advisor: P. K. Kitanidis.

Weber, Larry Joseph. A Study of the Flexural Properties of Alluvium Reinforced Ice Beams. M.S., 8/90. Advisor: W. A. Nixon.

Weber, Larry Joseph. A Study of Fracture Toughness and Fatigue of Freshwater Ice. Ph.D., 5/93. Advisor: W. A. Nixon.

Wei, Chi-Pang. No thesis. M.S., 5/75. Advisor: L. Landweber.

Whelan, Andrew Edward. Interfacial Fracture of Ice. Ph.D., 7/96. Advisor: W. A. Nixon.

Whelan, Gene. Distributed Model for Sediment Yield. M.S., 5/80. Advisors: S. C. Jain and T. E. Croley II.

Willis, Joe C. Sediment Discharge of Alluvial Streams Calculated from Bed-Form Statistics. Ph.D., 12/76. Advisor: J. F. Kennedy.

Woodhouse, Robert Allen. Evaluation of Thermal Standards on the Missouri and Upper Mississippi Rivers. M.S., 7/79. Advisors: A. R. Giaquinta and J. F. Kennedy.

Wright, Scott Alan. Thermal Regime of the Missouri River along Its Iowa Border. M.S., 5/97. Advisor: F. M. Holly, Jr.

Wu, Chun-Yen. An Experimental Investigation of Flexural Strength of Ice. M.S., 7/76. Advisors: C.-J. Chen and J. C. Tatinclaux.

Wu, Jay Jack. No thesis. M.S., 5/84. Advisor: C.-J. Chen.

Wung, Tzong-Shyan. Experimental and Numerical Study of Heat Transfer and Flow for Arrays of Tubes in Crossflow. Ph.D., 5/86. Advisor: C.-J. Chen.

Yang, Jinn-Chuang. Numerical Simulation of Bed Evolution in Multi-Channel River Systems. Ph.D., 12/86. Advisor: F. M. Holly, Jr.

Yang, Shih-An. Nonlinear Viscous Waves Produced by an Impulsively Moving Plate. Ph.D., 12/89. Advisor: A. T. Chwang.

Yean, Jungtsun. Force Exerted by Ice Sheet on Inclined Structure. M.S., 5/81. Advisor: J. C. Tatinclaux.

Yeh, Keh-Chia. Flow and Bed Topography in Fixed-Channel and River Bends. Ph.D., 12/90. Advisor: J. F. Kennedy.

Yeh, Rong-Chuen. Flow and Bed Topography in Meandering Alluvial Channels. M.S., 12/83. Advisor: A. J. Odgaard.

Yeh, Tso-Ping. Transverse Mixing of Heated Effluents in Open-Channel Flow. Ph.D., 5/74. Advisor: W. W. Sayre.

Yeh, Yeu-Pin. Finite Analytical Numerical Solution of Laminar Flow in Curved Ducts. M.S., 8/85. Advisors: H. C. Chen and C.-J. Chen.

Yildirim, Nevzat. Effect of Surface Layer on Free Surface Vortex. M.S., 7/78. Advisor: S. C. Jain.

Yildirim, Nevzat. Velocity and Temperature Fields in Long Sidearms of Cooling Ponds. Ph.D., 5/82. Advisor: S. C. Jain.

Ying, Ker-Jen. No thesis. M.S., 7/83. Advisor: A. J. Odgaard.

Yoo, Sungyul. Viscous Inviscid Interaction with Higher-Order Viscous-Flow Equations. Ph.D., 12/86. Advisors: F. Stern and V. C. Patel.

Yoon, Byungman. A Study of Atmospheric-Icing Formation and Forces. Ph.D., 5/91. Advisor: R. Ettema.

Yoon, Joon-Yong. Numerical Analysis of Flows in Channels with Sand Dunes and Ice Covers. Ph.D., 5/93. Advisor: V. C. Patel.

Zaphirakos, Alexander Nickolas. Operation of a Multipurpose Reservoir under Drought Conditions. M.S., 7/82. Advisor: P. K. Kitanidis.

Zhang, Zijie. Wave-Induced Separation. M.S., 8/95. Advisor: F. Stern.

Zhao, Hongbiao. Detection of Nonstationarity in Heavy Precipitation Series. M.S., 8/97. Advisor: A. A. Bradley.

Zufelt, Jon Eugene. Ice Jam Dynamics. Ph.D., 5/96. Advisor: R. Ettema.

Index